WHAT YOUR COLLEAGUES ARE SAYING...

Recognizing that students need to learn social and emotional skills is one thing; helping them do so is another. This book can help you do just that. With its easy-to-follow structure, The Well-Rounded Math Student *offers us ways to help students develop both intra- and interpersonal skills through specific instructional strategies situated within the context of mathematical activities, such as problem-solving, reasoning, argumentation, and precision, and brought to life with real classroom examples.*

—Peter Liljedahl
Professor of Mathematics Education,
Simon Fraser University
Vancouver, Canada

This book could revolutionize the world of education! It takes the concept of integrating interpersonal and intrapersonal life skills into the math classroom. It is informative and straightforward. The authors provide current, practical examples that help readers see the full vision of the author's work and how to implement the ideas quickly and effectively.

—Kali Swisher
School Counselor, USD 504
Oswego, KS

This book illuminates how to integrate the development of students intrapersonal and interpersonal skills alongside emphasizing the mathematical practices. The many classroom examples, dozens of competency builder ideas, and collection of logs and assessment tools equip teachers to explicitly attend to students social and emotional competencies, while also developing their mathematical proficiency.

—Jennifer Bay-Williams
Professor, Mathematics Education and Author
Louisville, KY

This book is exactly what teachers need right now. After working with the math practices for 20 years it's time to take them to a new level. Incorporating purposeful social-emotional learning into the math classroom will build connections between students and their math knowledge. It's a brilliant idea!

—Jill Solomon
Director of Academic Systems,
Oakwood School
North Hollywood, CA

Written by practitioners, for practitioners, the authors authentically demonstrate how to recognize and support students' social emotional development through mathematics planning and assessment. This resource seamlessly integrates social-emotional competencies with the eight Standards for Mathematical Practice. Shifting focus from the stigma of adding "more" to a teacher's plate, this book serves to encourage meaningful mathematical learning.

—Sara Schwerdtfeger
Dean, The Teachers College,
Emporia State University
Emporia, KS

The Well-Rounded Math Student *offers a practical guide for fostering strong math skills as well as social and emotional competencies. As a former elementary math coach, I've seen how building number sense, problem-solving abilities, and confidence in students leads to success. This book highlights the importance of creative thinking and self-regulation, providing strategies to nurture both academic and emotional growth in young learners. Ideal for teachers, administrators, and math advocates.*

—Antionette Stith
Assistant Professor of Education, Lake Superior State University
Marie, MI

The authors have thought deeply about our learners and their ever-evolving needs, as well as the math educators who are doing the brave work of teaching them on a day-to-day basis. The Well-Rounded Math Student: Leveraging Math Practices to Build Next Generation Skills *is written with both groups in mind. The work intentionally elevates competencies in a way that helps students access content, offering educators practical strategies that position their students as both capable mathematicians and thriving humans.*

—Tammy McMorrow
First-Grade Teacher,
Indian Creek Elementary
Kuna, ID

The Well-Rounded Math Student: Leveraging Math Practices to Build Next Generation Skills is a must-have resource for educators committed to whole-child development. By seamlessly embedding social-emotional competencies into mathematics instruction, this book equips teachers to foster self-efficacy, adaptability, and perseverance in their students. The practical tools, reflective questions, and real-world examples make it an invaluable guide for creating engaging, inclusive classrooms where academic and social-emotional growth go hand in hand. A transformative resource for educators and school counselors alike!

—Hanna Kemble Mick
Elementary School Counselor, Auburn-Washburn USD 437
Topeka, KS

This book bridges the gap between math instruction and social-emotional development, providing practical tools and insights I wish I had as a teacher. It is an extremely useful resource for math educators who want to cultivate not just problem solvers, but thoughtful, resilient, and adaptable learners.

—Kristin Wright
Research/Evaluation Associate, Office of Educational
Innovation and Evaluation (OEIE)
Kansas State University—College of Education
Clay Center, KS

If we want to build confident mathematicians who have self-efficacy and can talk fluently about math in a respectful classroom setting, this book walks us through just how easily we can achieve that! I feel empowered to make the shift in my lesson planning to meld social-emotional competencies with my math practices. It just makes sense! Make this your next professional development read!

—Stacey Johnston
Sixth-Grade Math and Science Teacher,
MEd Leadership, Central Middle School
Edmond, OK

This book challenged me to view the math classroom through a new lens, and after reading, I felt empowered to embed social-emotional competencies within math instruction in ways I had not thought of before!

—Kelly Oberheu
Former High School Mathematics Teacher
Emporia, KS

The Well-Rounded Math Student

The Well-Rounded Math Student

Leveraging Math Practices to Build Next Generation Skills

Sherri Martinie

Jessica Lane

Janet Stramel

Jolene Goodheart Peterson

Julie Thiele

CORWIN

FOR INFORMATION:

Corwin
A SAGE Company
2455 Teller Road
Thousand Oaks, California 91320
(800) 233-9936
www.corwin.com

SAGE Publications Ltd.
1 Oliver's Yard
55 City Road
London EC1Y 1SP
United Kingdom

SAGE Publications India Pvt. Ltd.
Unit No 323-333, Third Floor, F-Block
International Trade Tower Nehru Place
New Delhi 110 019
India

SAGE Publications Asia-Pacific Pte. Ltd.
18 Cross Street #10-10/11/12
China Square Central
Singapore 048423

Vice President and
 Editorial Director: Monica Eckman
Associate Director and
 Publisher, STEM: Erin Null
Senior Editorial Assistant: Nyle De Leon
Project Editor: Amy Schroller
Copy Editor: Heather Kerrigan
Typesetter: C&M Digitals (P) Ltd.
Proofreader: Dennis Webb
Cover Designer: Scott Van Atta
Marketing Manager: Margaret O'Connor

Copyright © 2026 by Corwin Press, Inc.

All rights reserved. Except as permitted by U.S. copyright law, no part of this work may be reproduced or distributed in any form or by any means, or stored in a database or retrieval system, without permission in writing from the publisher.

When forms and sample documents appearing in this work are intended for reproduction, they will be marked as such. Reproduction of their use is authorized for educational use by educators, local school sites, and/or noncommercial or nonprofit entities that have purchased the book.

All third-party trademarks referenced or depicted herein are included solely for the purpose of illustration and are the property of their respective owners. Reference to these trademarks in no way indicates any relationship with, or endorsement by, the trademark owner.

No AI training. Without in any way limiting the author's and publisher's exclusive rights under copyright, any use of this publication to "train" generative artificial intelligence (AI) or for other AI uses is expressly prohibited. The publisher reserves all rights to license uses of this publication for generative AI training or other AI uses.

Printed in the United States of America

ISBN 9781071949498

This book is printed on acid-free paper.

25 26 27 28 29 10 9 8 7 6 5 4 3 2 1

DISCLAIMER: This book may direct you to access third-party content via Web links, QR codes, or other scannable technologies, which are provided for your reference by the author(s). Corwin makes no guarantee that such third-party content will be available for your use and encourages you to review the terms and conditions of such third-party content. Corwin takes no responsibility and assumes no liability for your use of any third-party content, nor does Corwin approve, sponsor, endorse, verify, or certify such third-party content.

CONTENTS

List of Competency Builders   xvii
Preface   xix
    What This Book Is About   xix
    Who This Book Is For   xxi
    What To Expect From This Book   xxii
Acknowledgments   xxiii
About the Authors   xxvii

INTRODUCTION: The Role of Intrapersonal and Interpersonal Competencies in Mathematics Education   1
    A Paradigm Shift in Education   1
        The Shifting Goals of Education   2
        The Shifting Needs of Students and Demands on Schools   2
    Schools as a Hub for Whole Child Support   3
        The Shifting Role of the Teacher   5
    What Do We Really Mean by Social-Emotional Competencies?   6
        Intrapersonal Skills   8
        Interpersonal Skills   8
    Integrating Social-Emotional Competencies Into Our Mathematics Lessons   12
    Breaking Down the Process   14
        Beginning With Mathematical Content and Practice Standards   15
        Elevating the Inherent Social-Emotional Competencies   16
        Deciding on Instructional Structures and Engagement Strategies   17
        Assessing Progress Toward Mathematical Content, Practice, and Social-Emotional Goals   17
    The Lesson Planning Process   18
    Questions to Think About   20
    Actions to Take   20

CHAPTER 1: Building Self-Efficacy in Problem-Solving
for Mathematics and Daily Life   21

    Merging Content Standards, Mathematical Practices,
    and Social-Emotional Competencies   24

        Mathematical Content Standard and Corresponding
        Mathematics Goal   24

        Mathematical Practice   25

        Social-Emotional Competencies   26

            Intrapersonal Skills   27

            Interpersonal Skills   27

        Instructional Structures and Engagement Strategies   28

            Identifying a High-Quality Task   28

            Using Self-Efficacy Starters   31

            Seeing Mistakes as Opportunities to Learn   33

            Providing Constructive Feedback   34

            Connecting to Children's Literature   35

        Assessing Interpersonal and Intrapersonal Skills   35

    Looking at Exemplars in Action   38

        Ms. Patel and Their Kindergarten Class   38

        Mr. Nguyen and His Seventh Grade Class   40

        Reflection   42

    Summary   42

    Questions to Think About   43

    Actions to Take   44

CHAPTER 2: Fostering Self-Regulation and Sustained
Attention While Reasoning Abstractly and Quantitatively   45

    Merging Content Standards, Mathematical Practices,
    and Social-Emotional Competencies   47

        Mathematical Content Standard and Corresponding
        Mathematics Goal   48

        Mathematical Practice   48

        Social-Emotional Competencies   52

            Intrapersonal Skills   53

            Interpersonal Skills   53

Instructional Structures and Engagement Strategies	53
Using Self-Talk Strategies	54
Using Sense-Making Activities	56
Seeing Multiple Perspectives	61
Assessing Interpersonal and Intrapersonal Skills	67
Looking at Exemplars in Action	69
Mr. Kenyon and His Second-Grade Class	69
Mr. Thompson and His Sixth-Grade Class	72
Reflection	75
Summary	76
Questions to Think About	76
Actions to Take	77

CHAPTER 3: Growing Self-Awareness and Social Awareness Through Constructing and Critiquing Arguments — 79

Merging Content Standards, Mathematical Practices, and Social-Emotional Competencies	82
Mathematical Content Standard and Corresponding Mathematics Goal	82
Mathematical Practice	83
Social-Emotional Competencies	83
Intrapersonal Skills	84
Interpersonal Skills	84
Instructional Structures and Engagement Strategies	84
Defining Social Awareness	85
Constructing Arguments	86
Justifying Arguments	89
Critiquing Arguments	91
Assessing Interpersonal and Intrapersonal Skills	93
Looking at Exemplars in Action	96
Mr. Hernandez and His Fourth-Grade Class	96
Mx. Adams and Their High School Algebra Class	99
Reflection	102
Summary	102
Questions to Think About	103
Actions to Take	103

CHAPTER 4: Promoting Decision-Making When
Modeling With Mathematics   105

    Merging Content Standards, Mathematical Practices,
    and Social-Emotional Competencies   108

        Mathematical Content Standard and Corresponding
        Mathematics Goal   108

        Mathematical Practice   109

        Social-Emotional Competencies   111

            Intrapersonal Skills   112

            Interpersonal Skills   112

        Instructional Structures and Engagement Strategies   113

            Focusing on Decision-Making Approaches   114

            Using Physical Models to Explore Geometric Relationships   115

            Creating Opportunities to Ask Questions and
            Formulate Problems   116

            Using Children's Literature to Create Modeling Activities   121

        Assessing Interpersonal and Intrapersonal Skills   122

    Looking at Exemplars in Action   125

        Mr. Baker and His Fourth-Grade Class   125

        Mrs. Vaughn and Her High School Math Class   127

        Reflection   130

    Summary   130

    Questions to Think About   131

    Actions to Take   131

CHAPTER 5: Using Mathematical Tools to Build Adaptability
and Decision-Making Skills   133

    Merging Content Standards, Mathematical Practices, and
    Social-Emotional Competencies   135

        Mathematical Content Standard and Corresponding
        Mathematics Goal   136

        Mathematical Practice   136

        Social-Emotional Competencies   137

            Intrapersonal Skills   138

            Interpersonal Skills   138

Instructional Structures and Engagement Strategies	138
Teaching Students What Integrity Means	139
Planning With Intention	140
Giving a Textbook Task a Makeover	142
Setting Guidelines for the Use of Tools	145
Using Questioning to Engage Learners	146
Focusing on the Benefits and Limitations for Tools	147
Assessing Interpersonal and Intrapersonal Skills	148
Looking at Exemplars in Action	151
Mrs. Byrns and Her Second-Grade Class	151
Mr. Heinen and His High School Geometry Class	155
Reflection	157
Summary	158
Questions to Think About	158
Actions to Take	159

CHAPTER 6: Fostering Social-Awareness and Self-Efficacy Through Attending to Precision — 161

Merging Content Standards, Mathematical Practices, and Social-Emotional Competencies	163
Mathematical Content Standard and Corresponding Mathematics Goal	163
Mathematical Practice	164
Social-Emotional Competencies	165
Intrapersonal Skills	166
Interpersonal Skills	166
Instructional Structures and Engagement Strategies	166
Supporting Metacognition	167
Using a Graphic Organizer to Focus Attention on Important Aspects of Precision	168
Incorporating Worked Examples	169
Constructing Claim-Evidence-Reasoning (CER) Data Stories	171
Reflecting on Precision	173
Assessing Interpersonal and Intrapersonal Skills	175

Looking at Exemplars in Action	178
Mx. Santos and Their Third-Grade Class	178
Mrs. Sparkman and Her Seventh-Grade Class	180
Reflection	182
Summary	182
Questions to Think About	183
Actions to Take	184

CHAPTER 7: Leveraging Adaptability to Find and Use Structure — 185

Merging Content Standards, Mathematical Practices, and Social-Emotional Competencies	189
Mathematical Content Standard and Corresponding Mathematics Goal	189
Mathematical Practice	190
Social-Emotional Competencies	191
Intrapersonal Skills	191
Interpersonal Skills	191
Instructional Structures and Engagement Strategies	192
Finding Structure	192
Adapting Thinking	197
Assessing Interpersonal and Intrapersonal Skills	201
Looking at Exemplars in Action	203
Mrs. Reynolds and Her Fifth-Grade Class	203
Mr. Maier and His High School Algebra Class	207
Reflection	210
Summary	211
Questions to Think About	211
Actions to Take	212

CHAPTER 8: Increasing Perseverance by Exploring Repeated Reasoning — 213

Merging Content Standards, Mathematical Practices, and Social-Emotional Competencies	215
Mathematical Content Standard and Corresponding Mathematics Goal	215
Mathematical Practice	216

Social-Emotional Competencies	217
Intrapersonal Skills	217
Interpersonal Skills	217
Instructional Structures and Engagement Strategies	218
Fostering Curiosity	218
Teaching MP8 Explicitly	219
Using Children's Literature	220
Using MP8 to Meaningfully Teach Properties in Mathematics	221
Using Games	223
Assessing Interpersonal and Intrapersonal Skills	225
Looking at Exemplars in Action	227
Mrs. Sharp and Her First-Grade Class	227
Mrs. Slaman and Her Eighth-Grade Class	232
Reflection	236
Summary	236
Questions to Think About	237
Actions to Take	237

CHAPTER 9: From Awareness to Action: Final Thoughts on Merging Social-Emotional Competencies With Math Practices   239

Reflecting on the Outcomes of Implementing Competency Builders	239
Reviewing the Planning Process	240

References	**245**
Index	**253**

LIST OF COMPETENCY BUILDERS

CHAPTER 1

1.1   Teacher Task Analysis
1.2   I Used to . . . But Now . . .
1.3   Reflecting on Quotes
1.4   Learning From Mistakes
1.5   Analyzing Constructive Feedback
1.6   *Counting on Frank*

CHAPTER 2

2.1   Finding the Flipside
2.2   Estimation Exploration
2.3   Three Reads Routine
2.4   Information Gap
2.5   Multiple Representation Rotations
2.6   Graphic Organizers

CHAPTER 3

3.1   Brainstorming Social Awareness
3.2   Leveraging Concrete Representations
3.3   Notice and Choose Routines
3.4   What Makes Good Justification?
3.5   Conjectures and Counterexamples

CHAPTER 4

4.1   Identifying Problem-Solving Approaches
4.2   Building and Analyzing Shapes
4.3   Notice and Wonder Routines
4.4   Exploring Three-Act Math Tasks
4.5   Math Curse

CHAPTER 5

- 5.1 Fostering Core Values Associated With Integrity
- 5.2 Planning Questions
- 5.3 Comparing Tasks to Encourage Appropriate and Strategic Use of Tools
- 5.4 Implementing Guidelines for the Use of Tools
- 5.5 Questions to Support Appropriate and Strategic Use of Tools
- 5.6 Using Graphic Organizers for Tool Selection Evaluation

CHAPTER 6

- 6.1 Three Writes Routine
- 6.2 Four Corners Organizer
- 6.3 Worked Examples
- 6.4 CER Data Stories
- 6.5 Prompting Reflection

CHAPTER 7

- 7.1 What Do You Notice?
- 7.2 Card Sort
- 7.3 Cover and Uncover
- 7.4 Which One Doesn't Belong?
- 7.5 Math Talks

CHAPTER 8

- 8.1 Curiosity-Based Questions and Comments
- 8.2 How to Look for and Make Use of Structure
- 8.3 *When Sophie Thinks She Can't*
- 8.4 Introducing the Commutative Property
- 8.5 Introducing Properties of Exponents
- 8.6 Guess My Rule Game

PREFACE

AS A TEACHER SOMEWHERE IN THE P–12 TRAJECTORY, you know that your craft is both a science and an art. You know that students' knowledge, skills, and competencies grow as much through how you foster relationship, connection, and understanding (the art) as they do their academic learning (the science) (Bouffard, 2018). You likely joined the profession hoping to positively impact students' lives, just as another teacher once did for you. However, you also know impacting students' lives goes beyond delivering the content of a well-developed lesson. Research shows that learning social and emotional skills helps students academically (Benson, 2021; Institute of Education Sciences [IES], 2022; Jones et al., 2021; Langreo, 2023; Prinstein & Ethier, 2022). As P–12 students' needs continue to grow and change as they mature, teachers and staff become the frontline in social-emotional support, as most children and adolescents in the United States spend seven or more hours of their daily lives in school. This idea connects back to the core of teaching, which views teachers as trusted leaders who guide learning and use their knowledge, creativity, and connection in the classroom.

This book is intended to help you—as a teacher of mathematics, specifically—see how you can use what you already do, with some small shifts in intentionality, to foster both your students' academic prowess *and* their social and emotional development. It is designed to help you capitalize on all aspects of the student for a holistic approach to learning. This book will give you a new lens to consider and leverage in your planning as well as concrete ways to use your mathematics lessons to explicitly teach, highlight, and reinforce the social-emotional competencies—or the intrapersonal and interpersonal skills, sometimes referred to as Next Generation skills—your students need to build for ultimate success. As you will see, these competencies are already present (but sometimes unseen) within every P–12 mathematics lesson.

WHAT THIS BOOK IS ABOUT

Traditionally, teacher education programs from preschool through high school have focused exclusively on making sure teachers understand the subjects they teach and how to teach them. They have focused on the science of teaching through pedagogy, lifespan development, curriculum, and instruction. They have prepared teachers to learn how to create lessons and manage classrooms, keep students engaged, and adapt lessons for different needs (Tomlinson & Imbeau, 2023). Teaching content is important for student success, but it's only one part of what happens in classrooms today (Lane, 2018). While today's teachers are well-trained in teaching subjects like math, the challenges they face often go beyond this (Benson, 2021). Teachers must also know how to help students develop personal and interpersonal skills like self-reflection and communication.

The art of teaching, found and built within connections and relationships, is a key part of teaching. The student adage "I may not remember what you taught me, but I remember how you made me feel" still applies. Students may not remember the exact lesson or content standard, but they will remember that they learned and were supported as people. Teachers often innately rely on and draw from their own abilities to help, connect, and relate, but arguably the student-teacher relationship has been undervalued in today's education system. Additionally, intrapersonal and interpersonal skills have often been skills that teachers subconsciously assume students already have, and they develop lessons with that assumption. However, those student skills cannot be taken for granted and must be taught and developed alongside mathematical concepts and skills.

Over the past two decades, social-emotional learning (SEL) has surfaced as an overarching term for several concepts including character education, 21st-century skills, next generation skills, soft skills, employability skills, social skills, and trauma-informed learning (Jones et al., 2021). Regardless of the terminology used to encapsulate the skills we focus on in this book, the fundamentals of SEL states, "social-emotional learning (SEL) can help young people thrive personally and academically, develop and maintain positive relationships, become lifelong learners, and contribute to a more caring, just world" (Collaborative for Academic, Social, and Emotional Learning [CASEL], n.d.a). Social-emotional learning improves academic achievement (CASEL, n.d.b) and is a critical layer of prevention for children's mental wellness (Key Social Emotional Messages, CASEL, n.d.b). While teachers are in no way mental health providers, they are on the front lines of working with students and can offer and support preventive practices to promote student wellness.

> Social-emotional learning improves academic achievement and is a critical layer of prevention for children's mental wellness.

PREFACE

Intrapersonal and interpersonal skills—also referred to in this book as social-emotional competencies (SECs)—are an undervalued but necessary component in K–12 education and postsecondary life. Teachers teach and model these skills to help students use them in different situations, both in and out of the classroom. This helps make the classroom and school environment better and helps students have a more positive attitude about themselves, others, and school (Jones et al., 2021). Also, colleges and employers continue to seek evidence of intrapersonal and interpersonal skills, or next generation employability skills, such as clear communication, decision-making, problem-solving, and teamwork that are transferable to life beyond P–12 environments. These competencies are in high demand and valued just as much as academic and content knowledge.

The overarching goal of education has always been to help kids become responsible adults who are ready for their future. A teacher's impact in this comes from both academic rigor *and* through cultivating strong social and emotional skills that facilitate healthy relationships. These developed skills and relationships offer students the chance to grow into productive adults and citizens.

> The overarching goal of education has always been to help kids become responsible adults who are ready for their future.

WHO THIS BOOK IS FOR

This book is designed to help you reconnect with the art of teaching and support a holistic approach to developing and enriching critical thinking and employability skills to aid in content learning. As outlined, we know you are already highly skilled and trained in creating and executing lessons that support mathematical content and standards. This book is for any P–12 classroom teacher looking for a bridge to shape and reinforce intrapersonal and interpersonal skills within the mathematics classroom through enhanced lesson planning. This book is also for any preservice teacher seeking ideas on how to develop holistic lessons and coaches working with teachers to strengthen and enhance social development, connection, management, and community within classrooms.

WHAT TO EXPECT FROM THIS BOOK

As you begin this work, the Introduction will set the stage by describing modern shifts in education, including the goals of today's schools, today's students' needs, and the evolving and important role of the classroom teacher. From there, each chapter represents and reviews a mathematical practice and offers guided questions to help you consider each phase of a comprehensive lesson. You will gain a clear understanding of the inherent social-emotional competencies—both intrapersonal and interpersonal skills—connected to that practice that can be easily integrated and enhanced across the lesson.

You will capitalize on your strengths in content delivery to recognize and leverage the implicit connections between the standards for mathematical practice (MPs) and SECs. With a targeted lens and a few small tweaks to your current lessons, you will learn how to seamlessly draw out the intrapersonal and interpersonal components, making what is implicit clearer and more explicit, resulting in a more holistic lesson. After reading the book, using the competency builders, and reflecting on the narratives, you will

1. Feel comfortable identifying social-emotional competencies in your lessons.

2. Understand how, with a few small changes to your lesson plans, you can support students' intrapersonal and interpersonal growth along with their wellness during math lessons.

3. Feel confident and empowered to embrace this expanded way of thinking, uncovering the often-hidden social-emotional competencies in the classroom and math lessons. You already have the skills to incorporate a more holistic approach to learning, and after reading this book, you will be ready to make a difference in your very next lesson.

Our goal with this guide is to support your efforts in the classroom to teach and build social-emotional competencies without being burdensome. We know you are a highly trained professional, skilled in mathematical instruction, and we recognize and value that expertise. The demands on both you and your students are significant, and your role is crucial to their success. Your students rely on you, and you deserve practical solutions that deliver the greatest impact.

ACKNOWLEDGMENTS

From Sherri: To my children Curtis, Peter, and Lucy—your love and encouragement inspire me every day. Thank you for your patience and unwavering support. To my husband, Brian, I am thankful for your steady presence through every new endeavor, for enduring the highs and lows, and for celebrating each milestone with me. To my parents, I know you are with me in spirit, and I thank you for instilling in me a love of learning and solid work ethic. A heartfelt thank you to Jessica, Janet, Jolene, and Julie—your collaborations on this book has been invaluable. I truly enjoyed our time together! Finally, to my current and former colleagues and students, I am profoundly grateful. You have shaped my journey, challenged my thinking, and inspired me to keep learning and growing as an educator.

From Jessica: This book represents a full-circle moment—bringing together my passions and commitment to proactive, impactful work in education. I am profoundly grateful to my students, whose dedication and passion have deepened my love for counseling and teaching. Their commitment to creating better classrooms, schools, and communities continues to inspire me. The hard work they pour into P–12 schools does not go unnoticed, and this book is a tribute to the educators and school counselors who make a meaningful difference every day. I am also deeply appreciative of my colleagues, whose mentorship and encouragement have shaped my professional journey. I have been fortunate to work alongside and learn from some of the very best. Mary Napier, Libby Mellies, Megan Meile, Connie Carr, and Bobbi Murray strengthened my confidence and refined my skills in the P–12 classroom. Dr. Fred Bradley reinforced my belief in the vital role of relationships and interpersonal skills for school counselors, teachers, and students. Deb Andres and Dr. Tim Frey provided the space and support to integrate these skills naturally and authentically into teacher preparation courses. A special thank you to Dr. Sherri Martinie for recognizing the value of this work and inviting me to collaborate on this book. Sherri is not only a trusted colleague but also a mentor and friend. Finally, I extend my deepest gratitude to my children, Luke and Lauren. They are my greatest sources of strength and purpose. Balancing the demands of being both a counselor educator and a mother is challenging, and I am endlessly grateful for their patience, understanding, and unwavering support as I brought this book to life.

THE WELL-ROUNDED MATH STUDENT

From Janet: First and foremost, I want to express my heartfelt gratitude to Sherri for trusting in me and inviting me to be a part of this project. Your confidence in me and your friendship means more than words can convey. I am forever grateful to my husband, Dean, whose unwavering support and pride in my work have been a constant source of strength. My deepest appreciation also goes to my children—Tracy and Joshua, Luke and Anna—whose love and understanding have been a source of encouragement and joy throughout this journey. I am especially thankful for my parents, Bob and Norma Smith, whose pride in me and my work was always evident and deeply felt. Finally, to my sister, Kathy, who grew up alongside a "future teacher" long before I officially became one—thank you for being part of this journey from the very beginning. Your support has meant the world to me.

From Jolene: This book would not have been possible without the love, support, and encouragement of so many people who have shaped my journey. To my family—especially Josh, Izzie, Braeden, and Emmett—your love, patience, and unwavering belief in me have been my foundation. Thank you for always standing by me and for being my greatest joys. To the teachers who inspired me, challenged me, and nurtured my curiosity—you showed me the power of learning and the impact of a great educator. And to everyone who believed in me, even when I doubted myself—your faith in my potential gave me the courage to keep going. I am forever grateful.

From Julie: I would like to thank my family. I am blessed daily by my husband Shawn and our girls Tessa, Taryn, and Teagen. Your support throughout my career has kept me centered and focused on the importance of this journey and taking time for my faith, family, and friends while staying motivated to achieve my professional goals. To my parents, who instilled a love of learning by allowing me to explore the world around me and learn from my mistakes, who continue to provide encouragement, support, and guidance. I am thankful for my students, past and present, for encouraging me and allowing me to learn alongside them. I am grateful for the encouragement and strength I have received from my colleagues, teachers, professors, mentors, and friends. These amazing people are what make the journey fulfilling—thank you to all who have encouraged and even challenged me along the way, with whose support this book and my professional goals are possible.

PUBLISHER'S ACKNOWLEDGMENTS

Corwin gratefully acknowledges the contributions of the following reviewers:

Karla Bandemer
Grades 3–5 Math Teacher Leader, Lincoln Public Schools
Lincoln, NE

Casey McCormick
Grades 5–8 Math Teacher, Our Lady of the Assumption School
Citrus Heights, CA

Debbie Waggoner
Direct Instructional Coach, Fayette County Public Schools
Mathematics Methods Adjunct, Midway University
Lexington, KY

Tammy McMorrow
First-Grade Teacher, Indian Creek Elementary
Kuna, ID

Kelly Oberheu
Former High School Mathematics Teacher
Emporia, KS

Jill Solomon
Director of Academic Systems, Oakwood School
North Hollywood, CA

Stacey Johnston
Sixth-Grade Math and Science Teacher, M. Ed Leadership,
   Central Middle School
Edmond, OK

Kali Swisher
School Counselor, USD 504
Oswego, KS

ABOUT THE AUTHORS

Dr. Sherri Martinie, a professor of curriculum and instruction in the College of Education at Kansas State University, teaches undergraduate and graduate courses in mathematics education. Prior to taking her position at Kansas State, she taught elementary, middle, and high school mathematics for a combined 20 years. She is continually seeking innovative ways to support preservice and in-service teachers in the development and refinement of effective mathematics teaching practices.

Dr. Jessica Lane is an associate professor in counselor education and supervision at Kansas State University. Prior to serving as a counselor educator, she was an elementary teacher and school counselor in Kansas. She also served as faculty for nine years in preparing P–12 preservice teachers at Kansas State. Her research centers on prevention through school counseling, social-emotional support, rural counseling, and mental wellness.

Dr. Janet Stramel is a professor and Edna Shutts Williams Endowed Chair in the College of Education at Fort Hays State University (FHSU). She joined FHSU after teaching middle school math for 25 years. She currently teaches mathematics methods courses for preservice teachers. Her research focuses on STEM teaching and learning in rural schools.

THE WELL-ROUNDED MATH STUDENT

Jolene Goodheart Peterson is an education consultant for Smoky Hill Education Service Center as well as a teacher leader consultant for the Kansas State Department of Education. She has dedicated her educational career to mathematics, excelling as a teacher and instructional specialist. She focuses on effective teaching practices, fostering a mathematical mindset, and improving grading and reporting. Additionally, she is passionate about developing student character, service, and leadership.

Dr. Julie Thiele is an associate professor at Wichita State University. She teaches elementary mathematics methods, internship, assessment, and mentoring courses and serves as the instructional coordinator in the Teacher Apprentice Program. She leads professional development and conducts research with in-service and preservice teachers to enhance effective mathematics teaching and learning, as well as alternative certification pathways and online teacher preparation programs.

INTRODUCTION

The Role of Intrapersonal and Interpersonal Competencies in Mathematics Education

YOU MIGHT HEAR SOMEONE SAY, "When I was in school, we didn't need this (social-emotional support)" or "I've been teaching for years and we weren't trained to do these kinds of things, we just teach." However, today's schools look different than they did even one or two decades ago, and the insights on what is needed have mostly been ignored by the profession (Prinstein & Etheir, 2022). In this chapter, we will explore the current state of schools, the changing needs of students, and the shift in education necessary to support today's learners. While teachers are still key in delivering content, especially in math, the role now must include supporting students' social-emotional development and helping them to hone those next generation skills that will be critical into their adulthood.

Specifically, this chapter will

- Discuss the shift toward holistic teaching, which includes integrating social-emotional skills into lessons.
- Guide you step-by-step to enhancing your lesson plans with small changes that promote these skills, showing how careful preplanning and simple adjustments can make a big difference to student engagement and learning with minimal effort.

A PARADIGM SHIFT IN EDUCATION

The world is changing fast, with global forces like technological advances and economic growth impacting our daily lives and requiring new skills to succeed in the future. Instruction must now focus not just on traditional subjects like math and reading, but also on teaching students how to think critically, solve problems, and work well with others.

Because of these global shifts, how we teach must evolve to help students develop the necessary skills to meet the demands of a complex and ever-changing world. This requires us to consider fundamental shifts in education in three ways: the shifting goals of education, the shifting needs of students and demands on schools, and the shifting role of the teacher as schools have become more of a hub for whole child support.

The Shifting Goals of Education

To better understand where we have been and where we are going, it's important to briefly understand how K–12 education has shifted over the past 20–25 years. In 2001, under the academic mandate of No Child Left Behind (NCLB), schools nationwide were driven by academic initiatives and outcomes that focused heavily on assessments and accreditation requirements. Educational outcomes were measured by testing and assessment, placing heightened pressure and emphasis on teachers and students to succeed in these academic areas of focus. High-stakes testing in math and English language arts (ELA) became the primary metric for evaluating the academic success of students, teachers, and schools. With hyperfocus placed on math and ELA outcomes, adjustments were made within the school day for extended time and prioritized resources for assessed courses. Schools and districts adopted testing resources, hired instructional coaches, and developed supplemental class sections and seminars to meet this initiative's demands (Lane, 2018).

As this approach of academic accountability continued, some unintended consequences emerged. Under the pressures of high stakes testing, teachers began "teaching to the test" or focusing mainly on helping students pass these tests. This made students pay more attention to getting the right answers, instead of understanding the material. In elementary schools, recess, physical education, and art classes were shortened or put on a rotating schedule so there could be more time for math and reading. In high school, math and reading were seen as the most important subjects, and time for other subjects was reduced. As NCLB continued, many states and schools saw the inherent challenges of this academic mandate and the need for a broader definition of academic success.

The Shifting Needs of Students and Demands on Schools

While schools focused more on test scores to measure success, students' social, emotional, and developmental needs didn't go away. In fact, student well-being diminished, and mental health challenges increased during this time (Centers for Disease Control and Prevention [CDC], 2024). The CDC's (2021) Youth Risk Behavior Surveillance System reported that from 2009 to 2019, even before the COVID-19 pandemic, many students experienced emotional distress, such as feeling persistently sad and hopeless (CDC, 2021; Prinstein & Ethier, 2022). Nearly every group of young people reported poor mental health during this time. Alarming trends showed that one in five students thought about suicide, and about 1 in 11 tried to take their own life

(CDC, 2021; Prinstein & Ethier, 2022). Concerns of bullying, cyberbullying, and school safety also grew. Unfortunately, while these experiences and behaviors were trending up, they were also largely ignored by education and health care (Prinstein & Ethier, 2022).

The growing mental health needs and behavioral challenges became even worse in 2020 because of the COVID-19 pandemic. When COVID-19 hit in the spring of 2020, people globally were left isolated and uncertain within their homes. Schools closed, as did the rest of the world, and remained so for at least the first several months of the pandemic—some much longer. Throughout the academic year of 2020-2021, educators created a variety of make-shift school formats and structures to accommodate learning—from in-person full time but socially distanced classrooms, to part in-person/part online hybrid learning structures, to fully remote learning. Structures differed by state, by district, and sometimes even by building. As a result, nearly all formats of schooling were very different from what students, teachers, and families had been used to (Jones et al., 2021). Resources for student support also differed by state, by district, and by building. The changes caused by the pandemic impacted all students, but they were especially harmful to the most vulnerable groups. Social development was stifled by a lack of interaction with others, and in many cases "connection" occurred from behind a screen (Lane et al., 2020). For many, there was little to no connection at all. Isolation and heightened unknowns—including food insecurity, unstable income and housing, anxiety and fear wrought by a contagious disease, and dealing with family members' illness and death—created additional pressures on student well-being and mental health (Bonella et al., 2020). Chronic absenteeism has also increased post COVID-19 (Swanson et al., 2024), impacting classroom instruction, student interactions, and academic outcomes.

SCHOOLS AS A HUB FOR WHOLE CHILD SUPPORT

As schools began to reopen, most educators (82 percent) said their biggest worry returning to the classroom was the social-emotional well-being of students, even more than academic issues (Bonella et al., 2020). The National Center for Education Statistics (NCES, 2022) reported that "classroom disruptions from student misconduct (56 percent), rowdiness outside of the classroom (48 percent), and acts of disrespect towards teachers and staff (48 percent)" increased (para. 2). One in five educators felt unprepared to provide social-emotional support to students (Bonella et al., 2020).

In fact, 70 percent of public schools surveyed requested more training to help students with their social-emotional development (Institute of Education Sciences [IES], 2022). This concern was the same in rural, urban, and suburban schools.

The social impact of isolation and gaps in social-emotional development and academic learning created by COVID-19 impacts all aspects of student development. COVID-19 influenced an increase in depressive symptoms, anxiety, and stress in students (Zarowski et al., 2024). While short-term government funding was provided to help schools connect students with the mental health support they needed, efforts often fell short. The need for support was so great that schools found it difficult to identify and hire enough qualified school counselors to meet the demand. Availability of youth mental health professionals became impacted and wait times to see a therapist *outside* of school could be as long as six months.

Today, 84 percent of teachers believe that social-emotional learning has a positive impact on academic achievement (Bushweller, 2022; NCES, 2022). There is recognition that serious and ongoing mental health issues requiring professional intervention can't easily be fixed in a classroom and teachers are *not* expected to become formal mental health providers. However, preventive and supportive efforts focused on social-emotional learning *create* a protective factor for student's mental wellness. Protective factors, such as positive relationships, lower the likelihood of negative outcomes to enhance mental well-being. This means classroom teachers need to be part of the effort to build or rebuild social-emotional skills in students, creating strong and supportive relationships to negate stress, trauma, and other obstacles that students face (Jones et al., 2021).

In reviewing recent educational history assessing where we were and where we are, we now have a clearer picture of where we need to go moving forward. Simply put, solely focusing on academics is an antiquated way of teaching and learning, and educators must evolve and adjust to successfully meet today's needs. With this newfound understanding, for today's classroom teacher to be as successful as possible, there must be a paradigm shift toward a more holistic and preventive response. The way forward can be found in social-emotional competencies embedded within the classroom, and the synergy and strength that comes from addressing academic, social-emotional competencies to maximize student outcomes.

INTRODUCTION

> Solely focusing on academics is an antiquated way of teaching and learning, and educators must evolve and adjust to successfully meet today's needs.

The Shifting Role of the Teacher

Given the recent changes and challenges in education, it's clear that a new, integrated approach to teaching and learning is essential. While the needs are known, how exactly to bridge the gap between academic and social-emotional needs remains a concern. Many educators acknowledge that they were not trained in social-emotional skill development or delivery in their preparatory programs, and more instruction within this area of teaching is necessary, as "all professionals who work with young people need the knowledge to support students" (Abrams, 2023). We know teachers already feel stretched thin between higher student needs, delivering engaging lessons, and the burden of "doing more with less." The real question is, **"How do we do this?"**

How do teachers

- Lift up and enhance what they are already doing well?
- Capitalize on the relationships and connections already developed?
- Look for opportunities to tweak, not overhaul, their lessons to incorporate and highlight social-emotional competencies and support the whole child?

This book is intended to help you understand this. While challenges are evident, so are many assets. First, we begin by holding a strengths-based mindset and approach—for both ourselves as teachers and our students. Never has the care, concern, and impact of teachers been more needed. As caring and invested adults, we are our greatest resource for student success and serve as one of the largest influencers of success and learning (Jimerson & Haddock, 2015). As mentioned at the beginning of this book, many K–12 teachers join the profession hoping to positively impact students' lives, just as another teacher did for them. Grounded in purpose and driven by their work, teachers teach, inspire, and provide care, consistency, and stability within their classroom each day. You can likely relate to this.

Teachers already use social-emotional competencies within their lessons and classrooms every day; it now simply becomes a matter of intentionally and explicitly enhancing the already present intrapersonal and interpersonal skills found within the lesson. Through careful preplanning and enhanced, integrative lesson planning, we can teach and model desired skills and outcomes. Time spent creating strong lesson plans can help with engagement, classroom management, behavior, and learning and can positively impact students' social and emotional well-being.

Through small adjustments, teachers can encourage connectedness and productive social behavior while also teaching critical math concepts. This small adjustment can have great gains, and doesn't require more money, more time, or another trendy program that is not followed with fidelity.

WHAT DO WE REALLY MEAN BY SOCIAL-EMOTIONAL COMPETENCIES?

Over the past two decades, social-emotional learning (SEL) has surfaced as an overarching term for several concepts including character education, 21st-century skills, soft skills, employability skills, social skills, and trauma-informed learning (Jones et al., 2021). According to the Collaborative for Academic, Social, and Emotional Learning (CASEL, n.d.a), the fundamentals of SEL states, "social-emotional learning (SEL) can help young people thrive personally and academically, develop and maintain positive relationships, become lifelong learners, and contribute to a more caring, just world." Social-emotional learning improves academic achievement (CASEL, n.d.b) and is a critical layer of prevention for children's mental wellness (CASEL, n.d.b).

> Social-emotional learning improves academic achievement and is a critical layer of prevention for children's mental wellness.

Unfortunately, in recent years, some have tried to politicize and weaponize the term *social-emotional learning*. To be clear, regardless of packaging, social-emotional learning or competencies are not controversial (Prinstein & Ethier, 2022). Social-emotional learning are skills that help students better understand themselves and interact well with others. Teachers teach and model these skills to help students use them in different situations, both in and out of the classroom. This helps make the classroom and school environment better and helps students have a more positive attitude about themselves, others, and school (Jones et al., 2021).

For this book's purpose, we offer terminology to help name and frame social-emotional competencies (SECs). Having a common language across lessons, classrooms, and grade levels reinforces and strengthens positive social-emotional competencies and behaviors. This shared understanding clarifies and amplifies these skills and enables K–12 teachers to better understand, apply, and integrate those skills into lessons. To the authors, social-emotional competencies are an overarching construct for the targeted skills needed to build and enhance healthy relationships. In Figure i.1, the SECs are the overarching theme, and the competencies are further broken down into two key concepts of intrapersonal and interpersonal skills.

Figure i.1 • *Social-Emotional Competencies*

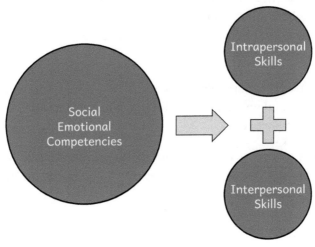

There are several bodies of work that describe social-emotional competencies, including the CASEL competencies framework and the College and Career Competency Framework and Wheel (https://www.cccframework.org). Not one of these descriptions quite fits the bill, so we have derived our shared language from these places as well as others.

Intrapersonal Skills

Intrapersonal skills have two distinct purposes:

1. **Highlighting the Relationship with Ourselves and the Learning Process**:
   - **Purpose**: Intrapersonal skills emphasize the significance of our internal relationship with ourselves. This means understanding our own thoughts, emotions, and motivations.
   - **Importance**: By developing intrapersonal skills, we become more aware of our own learning processes, which helps us better manage our personal growth and development. Essentially, these skills help us reflect on our own experiences and understand how we learn and grow.

2. **Understanding Ourselves to Better Interact with Others**:
   - **Purpose**: Intrapersonal skills also emphasize the need to gain self-awareness to improve our interactions with others.
   - **Importance**: By understanding our own thoughts, feelings, and behaviors, we become better equipped to relate to others. This self-awareness enables us to communicate more effectively, empathize with others, and work collaboratively. In other words, knowing ourselves well helps us understand and connect with people around us more effectively.

In summary, intrapersonal skills help improve our understanding of ourselves and how we learn and enhance our ability to interact and work with others by first gaining self-awareness.

We selected the following intrapersonal skills (Table i.1) based on their connection to the mathematical practices (MPs) and based on their importance for success in mathematics.

Table i.1 • *Intrapersonal Skills*

Competency	Description
Creative Thinking	Students develop unique and meaningful alternative ideas to solve the problem. They see the ideas from multiple perspectives and brainstorm ideas to explore using concrete referents such as objects, drawings, diagrams, and abstract approaches.
Curiosity	Students seek to learn or to know something for its own sake; desiring information to fill knowledge gaps and welcome new experiences. They look for connections or patterns, practice trial and error, and try new approaches to solve mathematical problems.

INTRODUCTION

Competency	Description
Goal Setting	Students establish objectives, monitor progress, and adjust strategies to achieve success. They develop a roadmap toward completing the mathematical problem within certain steps and a given timeframe.
Integrity	Students act with honesty, sincerity, fairness, and values. Teachers provide clear guidelines and high expectations, and students make strategic mathematical decisions that display academic honesty through the appropriate use of tools and strategies, and fairness with others.
Intrapersonal Communication (Self-Talk)	Students hold ongoing internal dialogue and reflective thinking by contextualizing a situation and considering how they would approach it. They use self-talk to think through and convince themselves before sharing an idea or making an argument.
Perseverance	Students persist and consistently work toward an outcome, even when difficult, until the result is achieved. They grapple with the mathematical process, working to determine the pattern or solve the problem.
Responsible Decision-Making	Students gather information, assess potential outcomes, and determine the best possible plan. They identify and consider options before making sound decisions about what steps to take, what tools would be useful, and if the outcome is reasonable.
Self-Awareness	Students hold a conscious understanding of one's own feelings, characteristics, thoughts, motivations, and desires. They are challenged to reflect on their feelings, thoughts, and values as they explore the truth of their conjectures and justify their conclusions.
Self-Efficacy	Students succeed in specific mathematical situations and tasks, which helps develop confidence in mathematical abilities. As students practice attending to precision and see improvements in their mathematical skills, they are likely to feel more confident in their ability to attempt future math problems.
Self-Regulation	Students manage thoughts, emotions, and behaviors during mathematical problem-solving, particularly in times of stress. They monitor their thoughts and emotions when they disagree with someone else's approach, and when giving and receiving feedback.
Sustained Attention	Students maintain focus and concentration over time. They engage in abstract and quantitative reasoning tasks, push through tedious processes such as decontextualization, looking for and generalizing patterns, working through the modeling process, attending to precision, and seeing the process through to the end.

Download this table at https://companion.corwin.com/courses/wellroundedmathstudent

We encourage you to actively incorporate and emphasize intrapersonal skills in your lessons whenever possible. These skills are often underdeveloped and less visible but are crucial to students' overall development. Prioritizing intrapersonal skills in lesson planning helps students build self-awareness and apply these skills effectively in various situations. Like all skills, they need practice to grow and improve, as intrapersonal skills directly impact interpersonal and cognitive skills.

Interpersonal Skills

Interpersonal skills largely focus on communication and relationships with others. These skills are important because they are a part of everything we do in math and are useful in many areas of life. The math classroom is a key place for students to use, develop, and sharpen interpersonal communication skills. The following interpersonal skills (Table i.2) were selected based on their connection to the MPs and their importance for success in mathematics.

Developing and strengthening communication skills and teamwork is recommended across all math practices. Effectively communicating mathematical ideas and strategies with others, working together, sharing strategies, and learning from each other's approaches to problem-solving naturally fosters teamwork and cooperation. For this book's purpose, we do not focus much time on these skills since they are so naturally embedded, but instead intentionally focus on other interpersonal skills to highlight how those might be implemented into a lesson.

While some skills are clearly either intrapersonal or interpersonal, other skills draw from both skill sets and fall somewhere in between. For example,

- **Communication Skills**: Often seen as more of an interpersonal skill, communication is how you interact with others and interpret their responses. However, communication also involves intrapersonal aspects, such as how you understand and manage your own messages and emotions.

- **Decision-Making Skills**: These can also involve both intrapersonal elements (e.g., personal judgment and self-awareness) and interpersonal elements (e.g., considering others' opinions and collaborating with them).

Table i.2 • *Interpersonal Skills*

Competency	Definition
Adaptability	Students adjust to new conditions or challenges with ease by changing to respond to new information or circumstances. They detect errors, explore consequences, and compare predictions, which allow the adjustment to the mathematical model, problem-solving process, or solution.
Assertiveness	Students clearly express wants, needs, and thoughts while respecting others, even when difficult. They impart their own thoughts, mathematical critiques, and ideas in respectful ways and are treated with respect when doing so, even if there is disagreement.
Communication	Students effectively exchange ideas, thoughts, and feelings between two or more people. They mathematically express themselves with clear and accurate expressions, interpretation of symbols, attention to detail, effective explanations, and contextual understanding.
Empathy	Students seek to understand, share, and respect the feelings of others. They show concern when responding to the arguments of others by trying to comprehend from another perspective whether they agree or not.
Social Awareness	Students listen to, reflect, respond, and empathize with others' experiences to understand social norms and other diverse perspectives. They consider the contexts and backgrounds that influence others' ideas, perspectives, and contributions.
Teamwork	Students discuss approaches, fairly contribute, respect other teammates, and reach consensus toward a shared goal. They work with others by engaging in shared work to arrive at a mathematical product or answer.

online resources ⬇ Download this table at https://companion.corwin.com/courses/wellroundedmathstudent

In summary, skills like communication and decision-making don't fit neatly into just one category. Instead, they lie on a continuum where they incorporate both intrapersonal and interpersonal elements. This means that these skills' development and application can be influenced by personal experiences and social interactions.

INTEGRATING SOCIAL-EMOTIONAL COMPETENCIES INTO OUR MATHEMATICS LESSONS

Integrating social-emotional competencies, or intrapersonal and interpersonal skills, into lessons means being aware of how to connect those skills to what we are teaching as important parts of learning. It's also important to note that, for many reasons, students may not possess these personal and social skills. Therefore, we as teachers must explicitly incorporate, teach, and scaffold these skills, even informally, to ensure their development.

In this book, we share a framework for planning a lesson that enables you to be intentional about amplifying SECs in your math classroom by integrating intrapersonal and interpersonal skills with the standards for mathematical practice. The goal is to help you build on what you already do naturally in your teaching by highlighting and making clear the intrapersonal and interpersonal connections and opportunities within the lesson. Lesson planning will be addressed in three parts: the standards for mathematical practice, social-emotional competencies, and the pedagogical decisions teachers make.

Each chapter is dedicated to one of the eight standards for mathematical practice. The chapter begins with the connection between specific content standards and the focus MP, as this is likely where you are most comfortable. Then, the overlap between the MP and SECs is provided and some intrapersonal and interpersonal skills that specifically engage in this MP are discussed.

Through each chapter is a framework of questions you can use as you plan a math lesson that merges content standards, MPs, and SECs—thereby resulting in a robust lesson plan template. To support your enhanced lesson plan development, the Well-Rounded Math Lesson Guide (Figure i.2) provides a comprehensive list of all the questions to consider as you develop your lesson.

Figure i.2 • *Well-Rounded Math Lesson Guide*

Topic:		Math Practices: *Which mathematical practice enhances understanding of this content standard?*
Content Standard:		
Math Goal: *What is the mathematical goal of the lesson?*		☐ MP1 Problem solve and persevere ☐ MP2 Reason abstractly and quantitatively ☐ MP3 Construct and critique arguments ☐ MP4 Model with math ☐ MP5 Use appropriate tools strategically ☐ MP6 Attend to precision ☐ MP7 Look for and make use of structure ☐ MP8 Look for and express patterns
Lesson Objective(s):		Intrapersonal skills: ☐ Creative thinking
Launch/Introduction	Intrapersonal / Interpersonal Skills: *What intrapersonal and interpersonal skills are inherent, are needed, and can be further developed for students while engaging in the MP? How will I explicitly address SECs?*	☐ Curiosity ☐ Decision-making ☐ Goal setting ☐ Integrity ☐ Perseverance ☐ Self-awareness ☐ Self-efficacy ☐ Self-regulation ☐ Self-talk ☐ Sustained attention

Lesson Activities: *With an eye on our math goal, how will I support social-emotional development as I engage learners in the MP? What structures, strategies, methods, and/or tools can I use?*	Intrapersonal/ Interpersonal Skills:	Interpersonal skills: ☐ Adaptability ☐ Assertiveness ☐ Communication ☐ Empathy ☐ Social Awareness ☐ Teamwork
Summary/Closing:	Intrapersonal / Interpersonal Skills:	
Assessment: *How will I assess students' progress toward the mathematical goal of this lesson, their engagement in the MP, and their ability to use and continue to apply SECs?* *How will I provide feedback?* *How will I build a chance for students to reflect on their skills?*	Intrapersonal / Interpersonal Skills:	

Download this figure at https://companion.corwin.com/courses/wellroundedmathstudent

BREAKING DOWN THE PROCESS

Successful lesson planning begins with purposeful preplanning. When developing holistic lessons that emphasize social, emotional, and cognitive learning, it is important to

1. Begin with the math content standards you will address.
2. Connect this with the mathematical practice outlined.

3. Determine the social-emotional competencies that naturally align.

4. Consider what is developmentally appropriate for the age group one is teaching.

5. Deliver the lesson and provide reflection to solidify learning by discussing the social-emotional skills used to meet the math components and looping around the lesson for reflection.

Beginning With Mathematical Content and Practice Standards

Regardless of location, today's modern mathematical standards in every state and province aim to develop mathematical concepts and skills more deeply and to engage learners more fully in thinking and doing mathematics than previous standards had done. All standards also carry with them some form of standards for mathematical practice. These are sometimes referred to as math practices, math processes, or process standards, and they apply broadly to K–12 students and describe the practices, habits, and expertise that characterize proficient mathematicians that we strive to develop in all students. They describe ways in which developing students can engage with mathematics in an increasingly more sophisticated manner as they mature, and their mathematical expertise grows throughout their education.

The first step of lesson planning is to ask,

- **What is the mathematical goal of the lesson?**
- **Which mathematical practice enhances understanding of this content standard?**

Each of this book's chapters begins by describing the overarching mathematical goal of the lesson and the corresponding math, followed by determining what mathematical practices stand out and/or what you want to emphasize. While there are opportunities to focus on several mathematical practices, being intentional about one practice supports the development of other mathematical practices because they are often intertwined with each other. Being intentional means planning for students to be engaged in mathematics in a specific way.

As you engage in the lesson planning process, this section will feel familiar and comfortable as this is how we have been trained to approach planning with a keen focus on the math content.

Elevating the Inherent Social-Emotional Competencies

Next, you will add to the traditional lesson plan by bringing the implicit intrapersonal and interpersonal skills to the forefront. Recall that in their simplest form, intrapersonal skills are knowing and regulating oneself, and interpersonal skills facilitate interaction with others (Glowiak & Mayfield, 2016). Remember, the addition of intrapersonal and interpersonal skills is not a change to the math goal, content, or standards, but a magnified lens to make the social-emotional competencies obvious. Math teaching is the foundation, and it becomes even stronger when combined with the natural social-emotional competencies that are part of the lesson.

Ask yourself,

- **What intrapersonal and interpersonal skills are inherent, are needed, and can be further developed for students while engaging in this MP?**
- **How will I explicitly address SECs?**

By consistently highlighting these skills, we help students reinforce concepts, connect cognitive and personal aspects, and encourage reflection in a seamless way. The opportunity to "name and frame" social-emotional competencies also emphasizes their interplay and importance within the lesson. Using shared language to identify these skills helps make them more intentional, familiar, and comfortable. This also gives students another way to connect with and understand the math content better. By drawing from, explicitly naming, and building on the social-emotional competencies embedded in a math lesson, we can shape, support, and complement them. We also know that some students have fears or beliefs that they are not strong or skilled in math. When leaning into and applying other skills (e.g., communicating, adaptability, perseverance) within a lesson, we can also support the student who may not feel as confident in learning math by encouraging them to implement intrapersonal and interpersonal skills to be resilient and successful with math.

> By drawing from, explicitly naming, and building on the social-emotional competencies embedded in a math lesson, we can shape, support, and complement them.

We begin by focusing on and determining intrapersonal skills and then identifying interpersonal skills. As you read the book, you'll find examples that highlight a few skills within lessons. This doesn't mean other skills aren't relevant; these are just a few concrete examples of *many* possibilities. When developing your lesson, remember that it's important to focus on one or two SECs at a time and teach them explicitly so students can recognize and develop them.

Deciding on Instructional Structures and Engagement Strategies

Shifting toward planning, we ask ourselves, **"With an eye on our math goal, how will I support social-emotional development as I engage learners in this math practice? What structures, strategies, methods, and/or tools can I use?"**

After determining the key elements of the lesson through a holistic view where we begin with the math, and supporting the social-emotional competencies, we can consider how to engage students in the learning by using specific instructional strategies.

Assessing Progress Toward Mathematical Content, Practice, and Social-Emotional Goals

As we move into assessment, ask yourself, **"How will I assess students' progress toward the mathematics goal of this lesson, their engagement in the mathematical practice standard, and their ability to use and continue to apply social-emotional competencies? How will I provide feedback? How will I build a chance for students to reflect on their intrapersonal and interpersonal skills?"**

While math assessment can be either formal or informal, social-emotional competency assessment is largely informal. Due to its often undervalued status, lack of standardization, teachers' limited exposure, and the developmental and growth-based application, it is best to assess students' use of SECs informally.

Having regular, informal, individual check-ins with students regarding intrapersonal and interpersonal skills is a good way for students to practice social skills, develop rapport and connection between student and teacher, and allows for the student to authentically share their experience through effective communication. Formative assessment allows for immediate feedback throughout the lesson. As we observe student discussions or interactions, we can quickly gather insight into what students understand

and help boost correct answers or redirect and support other answers. These are actions to build into the lesson plan.

Students need feedback and an opportunity to reflect on their awareness and development in using various social-emotional competencies. When we know ourselves, we are better able to work well with and help others. Self-reflection is a practice that we should embed into student awareness, to help them understand and regulate their emotions and actions in various settings. To establish assessment criteria, identify actions we want to promote or instill on the part of the student.

> Self-reflection is a practice that we should embed into student awareness, to help them understand and regulate their emotions and actions in various settings.

THE LESSON PLANNING PROCESS

Figure i.3 illustrates lesson development that purposefully integrates social-emotional skill development and cycles back to ensure SEC support. The process starts with the math content, including the content standards and the standards for mathematical practice, enhances the social-emotional competencies to build the lesson plan, and then plans for assessment. The arrows on the top of the figure indicate the reflection of and tying back to previous decision points to inform planning and strengthen connections. After completion of the lesson, the arrow below illustrates the reflective process necessary to determine the effectiveness of the lesson in the given areas of math and intrapersonal and interpersonal growth.

As teachers, the adjustment to traditional teaching preparation is in the *intentional planning and acknowledgment of SECs*, naming them in the lesson, allowing time to model and discuss them, as well as reflecting on how students engaged in these actions.

As you can see, there is not a significant change in content and lesson development, but an added emphasis on drawing forth the implied skills that many teachers infer students already possess. This small adjustment in front-end planning on social-emotional competencies and enhanced lesson reflection can lend itself to greater rewards in the lesson and the classroom. This is a small shift in the traditional way of developing lessons, but it has the potential to deliver large returns in social, emotional, and cognitive learning,

Figure i.3 • *Reflective and Holistic Planning and Teaching*

Math Content
- Math Standards
- Mathematical Practice

Connecting SEC to Math

Social Emotional Competencies (SECs)
- Intrapersonal
- Interpersonal

Reflecting and Connecting to SEC

Assessment
- Formative
- Summative

Holistic Lesson Reflection

and in ways that greatly benefit the learner and the teacher without being burdensome.

In the following chapters, we will focus on the process of applying this new approach to lesson planning by concentrating on the mathematical practice, the corresponding standards, and opportunities for amplifying the natural intrapersonal and interpersonal opportunities within a lesson.

Questions to Think About

1. In your own words, how do you define intrapersonal and interpersonal skills?
2. Consider lessons you already teach.
    a. How do you engage learners? What interpersonal activities come to mind?
    b. What reflective practices do you use? What intrapersonal activities come to mind?
3. How do you currently reinforce positive classroom behaviors?

Actions to Take

1. As you move through this book, consider some of your favorite lessons or favorite activities that you already enjoy teaching. By shifting your lens, you will likely notice in those lessons that you already have natural avenues for intrapersonal and interpersonal connections. Now consider how you might amplify or draw out those connections a bit more, by naming the skills and asking reflective questions to build a more holistic lesson that maximizes math content along with social-emotional competencies.

CHAPTER 1

BUILDING SELF-EFFICACY IN PROBLEM-SOLVING FOR MATHEMATICS AND DAILY LIFE

IN OUR DAILY LIVES, navigating decisions and solving problems is a fundamental part of being human. Children also encounter problems at every stage of their development, from learning to tie their shoes to understanding the intricacies of social interactions. As they grow, they naturally acquire tools and experiences to help them tackle these challenges. However, what if you could intentionally cultivate and hone these problem-solving skills in a way that empowers them for life?

In mathematics, you have the unique opportunity to do just that. By guiding students through the process of *making sense of problems and persevering in solving them*, you can equip them not just with mathematical strategies, but with a mindset that is invaluable beyond the classroom walls. It's about more than "just solving for x" or finding the area of a rectangle; it's about building a repertoire of strategies that students can draw on in any situation. This dual focus on gaining and retaining problem-solving techniques is what prepares students to confidently approach challenges, persist through difficulties, and succeed in making sense of problems—both in mathematics and in life. While Mathematical Practice 1 (MP1) outlines some key strategies for problem-solving in mathematics, these strategies are transferable to many nonmathematical situations.

> **MAKE SENSE OF PROBLEMS AND PERSEVERE IN SOLVING THEM**
>
> *Mathematically proficient students start by explaining to themselves the meaning of a problem and looking for entry points to its solution. They analyze givens, constraints, relationships, and goals. They make conjectures about the form and meaning of the solution and plan a solution pathway rather than simply jumping into a solution attempt. They consider analogous problems and try special cases and simpler forms of the original problem to gain insight into its solution. They monitor and evaluate their progress and change course if necessary. Older students might, depending on the context of the problem, transform algebraic expressions or change the viewing window on their graphing calculator to get the information they need. Mathematically proficient students can explain correspondences between equations, verbal descriptions, tables, and graphs or draw diagrams of important features and relationships, graph data, and search for regularity or trends. Younger students might rely on using concrete objects or pictures to help conceptualize and solve a problem. Mathematically proficient students check their answers to problems using a different method, and they continually ask themselves, "Does this make sense?" They can understand the approaches of others to solving complex problems and identify correspondences between different approaches. (CCSSM, 2010)*

Consider the actions associated with MP1 in Table 1.1. Review the actions listed in the first column and think about how they connect to mathematical problem-solving. Then, examine the second column, where these actions are interpreted in the context of everyday problem-solving scenarios, like accidentally locking your keys in the car, or realizing you're missing ingredients for a recipe you've started.

Table 1.1 • *Actions Associated With MP1 That Can Be Applied to Everyday Life*

In Mathematics	In Everyday Life
• Interpreting and understanding the problem before attempting to solve it	• Identify and understand the problem. What is happening?

In Mathematics	In Everyday Life
• Planning a solution pathway	• Generate viable solutions. What can I do about this?
• Monitoring progress and being willing to change one's approach if needed	• Evaluate the alternatives and decide. If this is not working, what else can I try?
• Connecting current situations to previously learned concepts	• Think about what you know about the alternatives. Has this happened to me before? Has this happened to anyone else that I know?
• Continually assessing the logic and coherence of one's approach alongside the approaches of others	• Weigh the cost and benefit of the alternatives. What makes the most sense? Implement a strategy and evaluate whether it is working. What other strategies might also work?

MP1 describes a comprehensive approach to mathematical problem-solving that emphasizes understanding the problem, considering multiple strategies, analyzing representations, engaging persistently with a positive mindset, verifying solutions, and reflecting on solutions with peers. Effective problem-solving requires determining what the problem asks, what's important, the best strategy, and whether the solution is reasonable. It also involves having the mindset to handle confusion and persist until a solution is found (O'Connell & SanGiovanni, 2013). MP1 consists of two key components, each with its own distinct attributes (Table 1.2).

Table 1.2 • Attributes Associated With MP1

When students make sense of problems, they	When students persevere in solving problems, they
• Understand what they are being asked and consider multiple strategies and tools to analyze the problem • Analyze the meaning of the problem • Actively engage in the problem • Ask if their answers make sense • Check answers using a different method	• Show patience and a positive attitude • Continue thinking about what they can learn from the problem

Download this table at https://companion.corwin.com/courses/wellroundedmathstudent

By incorporating MP1 into mathematics instruction, teachers can support students' social-emotional growth, including their **self-efficacy**, **perseverance**, and **social awareness**. As students engage with problems and work through challenges, they recognize their abilities and learn to persist. Self-efficacy, defined by Bandura (1995) as the belief in one's capacity to execute necessary behaviors, plays a key role in how students approach tasks and respond to obstacles. Teachers can enhance this process by intentionally integrating and modeling social-emotional skills throughout lessons, using strategies like discussing and reflecting on problem-solving actions. Students' **social awareness** grows as they learn to understand and appreciate diverse perspectives. By naming these skills and providing regular opportunities for reflection, teachers can create a supportive learning environment that fosters both academic and emotional development.

As with all mathematical practices, students engage differently based on age and experience. Regardless of the grade level, it is essential to plan deliberately for this practice. All lessons should actively (1) assist students in linking mathematics content with mathematics practices, (2) help students connect mathematics to their social-emotional development, and (3) reflect what is suitable for the age group they are designed for.

MERGING CONTENT STANDARDS, MATHEMATICAL PRACTICES, AND SOCIAL-EMOTIONAL COMPETENCIES

There are many decision points in planning mathematics lessons that explicitly incorporate MP1 and the relevant social-emotional competencies. In the introductory chapter, a framework was shared for building mathematics lessons with a social-emotional learning mindset. This framework does not require additional work but rather a thoughtful and intentional approach in the planning process. The framework consists of questions to prompt you to be purposeful in your lesson development, and shift thinking about the social-emotional competencies, so you are not adding "one more thing" but *the* thing that is foundational to deep and meaningful learning of mathematics.

Mathematical Content Standard and Corresponding Mathematics Goal

Start by asking, **"What is the mathematics goal of this lesson?"** For example, in kindergarten, students learn to "represent addition and subtraction with objects, fingers, mental images, drawings, sounds (e.g., claps), acting out situations, verbal explanations, expressions, or equations" (K.OA.A.1, CCSSM, 2010). In this standard, students experience addition and subtraction in a concrete context to develop an understanding of what it means to add and subtract, as well as solve addition and subtraction scenarios using different representations. One mathematics goal for a lesson focused on this standard would be for students to engage with addition and subtraction situations using various representations (visual, verbal, symbolic, contextual, and physical) to help them make sense of the operations and how they relate. This will enable us to deepen students' understanding of the concepts of addition and subtraction and to move beyond a focus on rote calculation to grasp the underlying meaning.

At the middle school level, students learn to "solve problems involving scale drawings of geometric figures, such as computing actual lengths and areas from a scale drawing and reproducing a scale drawing at a different scale" (7.G.A.1, CCSSM, 2010). One goal for a lesson addressing this standard would be for students to explore the concept of scale factor as the number of times one object's measure is multiplied to obtain a similar object's measure. They then use this understanding to find the dimensions of an actual object given a drawing of the object that has been reduced or enlarged by a certain amount (also called a scale). A floor plan of a house is an example of a scale drawing. Another goal would be for students to determine the scale factor between two figures. This standard strongly emphasizes the role of visualization. The ability to visualize and then represent geometric figures on paper is crucial for making sense of and solving geometric problems.

Mathematical Practice

After examining the mathematical goals within the standards, consider how they align with the mathematical practices. Ask, **"Which mathematical practice enhances understanding of this content standard?"**

The standards emphasize using multiple representations (visual, verbal, symbolic, contextual, and physical) to help students understand concepts and solve problems. These standards encourage students not only to compute but also to analyze, identify relationships between representations, and make sense of problems. This approach fosters focus, patience, and adaptability as students plan, monitor, and, if needed, revise their strategies. These expectations align well with MP1. For instance,

- Use addition and subtraction within 20 to solve word problems involving situations of adding to, taking from, putting together, taking apart, and comparing, with unknowns in all positions, e.g., by using objects, drawings, and equations with a symbol for the unknown number to represent the problem. (1.OA.A.2, CCSSM, 2010)

- Tell and write time to the nearest minute and measure time intervals in minutes. Solve word problems involving addition and subtraction of time intervals in minutes, e.g., by representing the problem on a number line diagram. (3.MD.A.1, CCSSM, 2010)

- Explain each step in solving a simple equation as following from the equality of numbers asserted at the previous step, starting from the assumption that the original equation has a solution. (HS.A.REI.1, CCSSM, 2010)

These examples highlight not only the importance of problem-solving and multi-step problems but also the ability to represent problems in various ways, a crucial element of mathematical proficiency. By requiring students to use objects, drawings, and equations to solve addition and subtraction problems (1.OA.A.2), solve problems involving time (3.MD.A.1), or solve equations (HS.A.REI.1), students develop a deeper understanding of the underlying concepts. This approach supports the idea that there are often multiple pathways to a solution, encouraging students to explore different strategies and choose the most efficient one.

By encouraging students to understand a problem, devise a plan, and persist in finding a solution, these standards foster deeper mathematical understanding. Guide students in verifying their solutions by asking, "Does this make sense?" Across all grade levels, students are motivated to explore various strategies and persist through challenges.

Social-Emotional Competencies

Next ask, **"What intrapersonal and interpersonal skills are inherent, are needed, and can be further developed, while students engage in making sense of problems and persevering in solving them?"** As mentioned in the Introduction, integrating intrapersonal and interpersonal skills into lessons requires being mindful and intentional, so these skills are seen as a natural part of learning, not separate from the content. Students come with different levels of these skills, therefore it's important to explicitly teach and support them, even informally, to help students develop them fully. To accomplish this, pinpoint the intrapersonal and interpersonal skills that naturally align with the mathematical tasks students are completing. Although many of these skills can be applied during MP1, it's important to focus on one or two at a time, teaching or emphasizing them explicitly. This approach helps students recognize when they are using these skills and encourages their ongoing development. To illustrate this, let's focus on the intertwined intrapersonal skills of **self-efficacy** and **perseverance** and the interpersonal skill of **social awareness**.

Intrapersonal Skills

As mentioned at the beginning of this chapter, **self-efficacy** is the belief in one's ability to succeed in specific situations or accomplish a task. MP1 encourages students to reflect on their problem-solving approaches and consider alternative strategies. This process of reflection and adjustment demands perseverance, as students must be willing to continue working on a problem even if their initial approach doesn't yield immediate results. This idea pairs with Carol Dweck's work, which recognizes that students with a growth mindset, who believe in their capacity to improve through effort, are more likely to **persevere** when faced with challenges. **Self-efficacy** and **perseverance** are therefore mutually reinforcing, and MP1 cultivates this mindset of nimbleness and determination in mathematical problem-solving.

Interpersonal Skills

As noted in the introductory chapter developing and strengthening **communication skills** is recommended across all MPs. In the case of MP1, making sense of problems often involves discussing mathematical ideas and strategies with others. This includes listening to others' arguments, showing an understanding and respect for their ideas, and considering the various contexts and backgrounds that shape their perspectives. To promote this, create lessons that emphasize collaboration and discourse. Doing so also supports the development of **social awareness**, which allows students to appreciate diverse ways of thinking and helps create a supportive learning environment where all ideas are valued. When students collaborate with partners or in groups, MP1 helps them develop skills like listening to others, understanding different approaches, and respecting diverse perspectives. Engaging with others' ideas might present disagreements or conflicting perspectives, and **perseverance** in this context means staying engaged in discussions, listening carefully, and being open to new ideas, even if they seem difficult at first.

Instructional Structures and Engagement Strategies

Now shift toward planning and ask, **"With an eye on our mathematics goal, how will I support social-emotional development as I engage learners in MP1: Make sense of problems and persevere in solving them? What structures, strategies, methods, and/or tools can I use?"**

There are two important aspects of this MP to consider as you plan to implement it in the classroom. First, what approach to problem-solving are you using and how can you center the reasoning processes around student thinking? Second, how do you build self-efficacy in a way that supports perseverance in problem-solving? The teacher and student activities that follow speak to these considerations and demonstrate how to address these competencies explicitly.

Identifying a High-Quality Task

High-quality mathematics tasks have two critical components: the what and the how. Traditional word problems are simple and have one correct answer. Students are often taught to solve them by finding keywords and numbers and then calculating the answer; the perceived goal for the students becomes simply to arrive at the correct answer. This approach—sometimes called teaching *for* problem-solving—deprioritizes the thinking process. Conversely, a high-quality mathematics task is open-ended and/or can be solved in more than one way. High-quality tasks generate discussion, questioning, and critical thinking. They allow students to understand the context, explore various methods for solving tasks, and eventually decide on an efficient strategy to solve the problem. This approach is thought of as teaching *through* problem-solving and is especially important to building conceptual understanding. To support the selection of high-quality tasks, SanGiovanni (2017) created an Identifying High-Quality Tasks rating tool (Figure 1.1) that can be used to evaluate the quality and effectiveness of mathematical tasks.

Figure 1.1 • *Identifying a High-Quality Task Rating Tool*

Identifying a High-Quality Task

The purpose of the task is to teach or assess:

☐ Conceptual understanding ☐ Procedural skill or fluency ☐ Application

Rating:

2 – Meets the Characteristic

1 – Partially Meets the Characteristic

0 – Does Not Meet the Characteristic

The mathematics task:	Rating
Aligns to the content standards	
Promotes deep understanding of mathematics concepts rather than procedural knowledge	
Is relevant and engaging for students	
Has multiple entry points (self-efficacy)	
Encourages students to make sense of problems and persevere in solving them (perseverance)	
Encourages students to be actively engaged in discussions (social awareness)	
Allows for different strategies for finding solutions (adaptability)	

Source: From SanGiovanni (2017, p. 2). Reprinted with permission.

 Download this figure at https://companion.corwin.com/courses/wellrounded mathstudent

John SanGiovanni also developed a corresponding rubric that describes the way a task may or may not meet the intended characteristic as summarized in Figure 1.2. The rubric helps you evaluate the following criteria on a scale of fully meets, partially meets, and does not meet.

Figure 1.2 • *Task Characteristic Rating Rubric*

The task aligns to the mathematics standards I am teaching.
Tasks must be worthwhile and aligned to the skills and concepts in our curriculcum.
The task encourages my students to use representations.
Representations help students make sense of and communicate mathematical ideas.
The task provides my students with an opportunity for communicating their reasoning.
Students can communicate their reasoning with models or pictures, numbers, and words.
The task has multiple entry points.
Students can approach a problem from various perspectives, using diverse strategies and/or representations.
The task allows for different strategies for finding solutions.
Students can solve a problem in various ways.
The task makes connections between mathematical concepts.
Mathematics ideas are related. We can also connect them to representations, procedures, and applications.
The task prompts cognitive effort.
High quality tasks should generate some amount of struggle. Students should have to make sense of the prompt, the problem, or the representation.
Tasks are problem based, authentic, or interesting.
High-quality tasks are problem based. They can reflect real-world, authentic applications of mathematics. They should have interesting or noval prompts that grab students' attention.

You can download the full rubric at https://companion.corwin.com/courses/wellroundedmathstudent.

Notice within the descriptions of the characteristics are connections to social-emotional learning saying things, such as "High quality tasks should generate some amount of struggle, " Students make sense of and communicate mathematical ideas," and "Students can approach a problem from various perspectives."

> **Competency Builder 1.1**
>
> *Teacher Task Analysis*
>
> Use the Identifying a High-Quality Task Rating Tool (Figure 1.1) and the corresponding Task Characteristic Rating Rubric (Figure 1.2) to analyze the tasks you have selected for your next lesson. Next, consider modifications that

could improve the quality, accessibility, and rigor of each task. Then, review sources with examples of high-quality tasks. Look for opportunities to replace or enhance tasks in your existing materials with those that foster greater student engagement and deeper thinking.

When you use the Identifying a High-Quality Task Rating Tool and the Task Characteristic Rating Rubric, the social-emotional competencies are more explicit. **Self-efficacy** grows through tasks that offer appropriate challenges, and social awareness is enhanced through collaboration.

Some Sources for High-Quality Tasks

Illustrative Mathematics (https://tasks.illustrativemathematics.org/content-standards)

Youcubed (https://www.youcubed.org/tasks/)

Illuminations (https://illuminations.nctm.org)

3 Act Math Tasks (https://tapintoteenminds.com/3act-math/)

Rich Math Tasks (https://us.corwin.com/landing-pages/rich-math-tasks)

Using Self-Efficacy Starters

As students tackle difficult problems, you can attend to the development of their self-efficacy through several concrete skill building activities so that they can build perseverance. The following two activities illustrate each approach, and by doing both activities with students, you are doing this work explicitly, in small doses, applied consistently over time. You are giving students opportunities to understand perseverance and act themselves into the belief that they are capable of doing mathematics.

Competency Builder 1.2

I Used to . . . But Now . . .

Begin by defining self-efficacy and persistence with your students (or reviewing the definition if it has already been introduced). Share that self-efficacy is believing you can do hard things, and persistence is continuing to do something even when it is hard. Ultimately, there are two parts, believing and doing, but not necessarily in that order. Explain that it is normal to think that something

(Continued)

(Continued)

might be hard, and to question your abilities, but you are constantly growing, and you will know more and be able to do more tomorrow than you can today if you make the effort to learn. Over time, that effort adds up to a lot of learning and something that seems "too hard" now will not be hard in the future.

Ask students to name something that seemed hard to learn that they can do now with ease. Look outside the mathematics classroom for examples, such as tying shoelaces, riding a bike, running a mile, or beating a certain level of a video game. Throughout the year, students reflect on their learning and growth and respond to the prompt, "I used to think . . ., but now I . . . ." For example, "I used to think subtraction was hard, but now I know how to think about addition to help me solve subtraction problems" or "I used to think solving two-step equations was hard, but now I know how to break them down into manageable steps and solve them with confidence."

Competency Builder 1.3
Reflecting on Quotes

Share a quote and ask students to think about what the quote means and then share with a partner. Ask them to think about how it relates to self-efficacy and persistence (believing you can do hard things and continuing to try) and discuss as a class. Finally, ask them to think about how this applies to learning math. Here are some quotes to get you started.

- You are off to great places, today is your day. Your mountain is waiting so get on your way. —Dr. Seuss
- It doesn't matter how slowly you go if you don't stop. —Confucius
- You must expect great things of yourself before you can do them. —Michael Jordan
- Hard days are the best because that's when champions are made. —Gabby Douglas
- There will be obstacles. There will be doubters. There will be mistakes. But with hard work . . . there are no limits. —Michael Phelps
- If plan A isn't working, I have plan B, plan C, and even plan D. —Serena Williams
- You have to be able to accept failure to get better. —LeBron James
- Do not go where the path may lead, go instead where there is no path and leave a trail. —Ralph Waldo Emerson

CHAPTER 1 • BUILDING SELF-EFFICACY IN PROBLEM-SOLVING

Seeing Mistakes as Opportunities to Learn

Encourage students to recognize and reflect on their mistakes, as this helps them see how much they've learned and builds perseverance through the "power of yet" (Dweck, 2006). This mindset transforms challenges into growth opportunities, encouraging students to keep trying and believe in their ability to succeed with effort.

Ask students,

- Are you really learning if you are doing something you already know how to do?
- How do you know when you are learning something new?
- Yes! It is hard! You will do it wrong many times before you get it right. That is why mistakes are so important to learning!
- You may not have the answer *yet*, but I know you will keep trying!

Mistakes also help us target specific areas to focus on and practice and identify challenges in the first step toward setting goals and learning new things.

Competency Builder 1.4
Learning From Mistakes

Have students engage in a reflection activity where they identify a mistake they made (personal or academic) and either explain or draw the mistake. Then, have them explain or draw what they learned from that mistake. Students can keep this in a journal or a log (Table 1.3).

Table 1.3 • *Mistakes That Become Learning Log*

Mistake	Lesson Learned

online resources: Download this table at https://companion.corwin.com/courses/wellrounded mathstudent

Embed these strategies into problem-solving tasks that students encounter. Brief interventions that target students' mindset toward mistakes and their ability to learn new things have been shown to impact perseverance (Marshall, 2017).

Providing Constructive Feedback

Providing constructive feedback encourages students to persevere by acknowledging their efforts in tackling difficult problems and guiding them toward strategies that help them overcome obstacles. This improves their mathematics skills and social-emotional competencies. There has been a movement to provide better feedback, but what we are working on is getting students to respond appropriately to the feedback.

> **Competency Builder 1.5**
> *Analyzing Constructive Feedback*
>
> To build self-efficacy and social awareness, create structured opportunities for students to be mindful of the feedback they receive and to respond constructively. Provide time, space, and a log (Table 1.4) for recording feedback and their emotional and behavioral responses. Encourage students to note physical sensations and emotions triggered by feedback and to observe if they react defensively or with curiosity. Reflecting on feedback helps students understand their reactions, especially if they have challenging experiences in subjects like math, and fosters clarity and openness to learning, showing growth over time.
>
> **Table 1.4** • *Student Log to Record Response to Feedback*
>
> | Feedback | How I Felt | How I Acted | Benefits |
> |---|---|---|---|
> | | | | |
> | | | | |
> | | | | |
> | | | | |
>
> Download this table at https://companion.corwin.com/courses/wellroundedmathstudent

Connecting to Children's Literature

Children's literature offers students an opportunity to exercise their problem-solving skills as they relate to characters in stories and engage with the challenges they face. Many children's books offer a mathematics context.

> ### Competency Builder 1.6
> *Counting on Frank*
>
> Consider the book *Counting on Frank* by Rod Clements. It follows a dog named Frank and his imaginative owner as they present intriguing mathematical claims that involve counting, measuring, estimating, and calculating.
>
> After reading aloud the book, ask students to share all the ways the boy in the story "uses his brain." He uses it to ask questions, create problems, collect information, make calculations, and draw conclusions. Next, have students discuss with a partner how using the mathematical practices of making sense of problems and persevering in solving them enabled the boy to build self-efficacy. Go back to the page where he calculates that 24 Franks could fit into his bedroom. Tell the students they will now use their brains to determine how many Franks could fit into the classroom. To make sense of and solve this problem, students will need to ask for more information. For example, "How big is Frank?" Frank weighs about 100 pounds and fits into a large dog crate. They will need to gather more information (e.g., the dimensions of a large dog crate and the dimensions of the classroom). Solutions will vary depending on how students interpret and approach the problem. As they share, encourage them to reflect on the process they used and what they might do differently next time.

Assessing Interpersonal and Intrapersonal Skills

Last, ask, **"How will I assess students' progress toward the mathematics goal of this lesson, their engagement in the mathematical practice standard, and their ability to use and continue to develop social-emotional competencies? How will I provide feedback? How will I build in an opportunity for students to reflect on the development of the social-emotional competencies?"**

Recall the attributes in Table 1.2 at the beginning of the chapter. When observing student discussions, look for evidence of students making sense of problems and persevering in solving them. Use a whole class or individual observation tool (Tables 1.5 and 1.6) to document their development in mathematics content knowledge, engagement in MP1, and intrapersonal skills. These same prompts can be given to students to help them reflect on their own learning.

- How did you demonstrate **perseverance** today? Provide an example.
- How did you practice **self-efficacy** in today's mathematics activities? Provide an example.
- Share two examples of your **social awareness**. Were these interactions positive or negative? How could you improve?

Encouraging self-reflection helps students connect mathematical practices with social-emotional skills, fostering better self-awareness, collaboration, and emotional regulation. Use Table 1.7 to support self-assessment of these skills.

Table 1.5 • *Whole Class Observation Tool*

Name	Mathematics Goal	Engagement in Practice Standard *Make sense of problems*	Engagement in Practice Standard *Persevere*	Intrapersonal Competency *Self-efficacy*	Interpersonal Competency *Communication skills*

Note: *Progress will be marked using 0–No evidence, 1–Little evidence, 2–Adequate evidence*

 Download this table at https://companion.corwin.com/courses/wellroundedmathstudent

Table 1.6 • *Individual Student Observation Tool*

Name of student:			
Mathematics Goal	**No Evidence**	**Some Evidence**	**Adequate Evidence**
Engagement in the Mathematical Practice	**No Evidence**	**Some Evidence**	**Adequate Evidence**
Sense-making			
Perseverance			
Social-Emotional Competencies	**No Evidence**	**Some Evidence**	**Adequate Evidence**
Self-efficacy			
Communication			

Download this table at https://companion.corwin.com/courses/wellroundedmathstudent

Table 1.7 • *Self-Assessment Checklist*

Social-Emotional Competency	Not Sure	Not Yet	Getting There	Got It
Sense-making				
Perseverance				
Self-efficacy				
Communication				
Other skills used:				

Download this table at https://companion.corwin.com/courses/wellroundedmathstudent

LOOKING AT EXEMPLARS IN ACTION

Now that the merging of content standards, mathematical practice standards, and social-emotional competencies has been explored, let's look more closely at an elementary and a secondary example, using the standards and competencies focused on in this chapter.

Ms. Patel and Their Kindergarten Class

Ms. Patel's kindergarten classroom is focused on the mathematics standard "represent addition and subtraction with objects, fingers, mental images, drawings, sounds (e.g., claps), acting out situations, verbal explanations, expressions, or equations" (K.OA.A.1, CCSSM, 2010). As you read the vignette from this kindergarten classroom, identify how the teacher engages the students in MP1: Make sense of problems and persevere in solving them while helping them develop self-efficacy, perseverance, and social awareness. Think about ways you do this in your own classroom.

> Ms. Patel gathered the kindergarten children on the carpet, the center of which contained a basket of colorful counters. Ms. Patel begins, "Today, we are going to learn different ways to add and subtract using these counters." As they hold up a red counter and a yellow counter, they ask, "Who can tell me what happens when we put these two counters together?"
>
> Several children are excited to answer Ms. Patel's question. Miguel responds, "They make 2!"
>
> "Yes Miguel!" Ms. Patel says with a smile. "Now let's see if we can find different ways to make 4 using our counters."
>
> Ms. Patel asks two other students, Addison and Brennan, to demonstrate making 4.
>
> Addison chooses 2 red counters and 2 yellow counters and places them in a row. "2 and 2 make 4!" Addison says excitedly.
>
> Brennan places 1 red counter and 3 yellow counters on the carpet. "1 and 3 make 4 too!" she adds.

Ms. Patel then asks the students to work in pairs and explore making 4 with their own counters. Samantha and Miguel sit together, taking turns creating different combinations. Miguel says, "Look, we can make 4 this way too!" as he arranges the counters in a new pattern.

Ms. Patel walks about the room, kneeling to talk with each of the children as they work. They listen to their conversations, asking questions to guide their thinking. "Good job of finding 4. Now, can you show me *another* way to make 4?" they ask one group.

When students encounter challenges, Ms. Patel encourages them to persevere. "It's okay if you are stuck for a moment. I like how you keep trying different things, that's showing perseverance. Remember when we read the book, *The Little Engine That Could*?"

Ferdie enthusiastically says, "That was my favorite story!" Ms. Patel asks Ferdie why that one was his favorite. "The little engine kept trying and kept saying 'I think I can, I think I can, I think I can.' He kept trying to get over the mountain," replies Ferdie.

"That's right," says Ms. Patel. "Keep trying and believing you can solve it. Let me know if you'd like to think through it together," they say.

As a closing to the lesson, Ms. Patel gathers the children back on the carpet. "Today, we learned that there are many ways to add and subtract," they say. "You can use counters, your fingers, or even your imagination! Remember, it is important to make sense of problems and keep trying and believing in yourself, just like you did today." ●

Ms. Patel used hands-on activities with colored counters to support students' understanding of addition and subtraction. By asking questions like "what happens when . . ." they encouraged students to explore different methods and explain their thinking to each other, fostering both self-efficacy and interpersonal skills. As students encountered challenges, they reinforced perseverance by acknowledging their efforts and motivating them to keep trying. Through structured activities and social interactions, students learned to collaborate, share ideas, and adjust to peers' perspectives, enhancing their social awareness. Ms. Patel's approach nurtured self-efficacy, perseverance, and social awareness in their students.

Mr. Nguyen and His Seventh-Grade Class

Mr. Nguyen is emphasizing the mathematics standard "solve problems involving scale drawings of geometric figures, such as computing actual lengths and areas from a scale drawing and reproducing a scale drawing at a different scale" (7.G.A.1, CCSSM, 2010). As you read the classroom vignette, identify how the teacher engages the students MP1: Make sense of problems and persevere in solving them while helping them develop self-efficacy, perseverance, social awareness, and adaptability. Think about ways you do this in your own classroom.

> Mr. Nguyen is ready to introduce the seventh-grade lesson on scale drawings. He begins the lesson by discussing what scale drawings are and why they are used in real life. He asks the class where they have seen scale drawings. Renee says enthusiastically, "Maps!" Latisha responds, "And floor plans."
>
> Mr. Nguyen projects onto the whiteboard a scale drawing of a local park in which 1 inch represents 50 feet and explains, "Today, we are going to solve problems involving scale drawings of geometric figures."
>
> Mr. Nguyen calls on two students, Matthew and Amayah. He points to the distance on the scale drawing and asks, "According to the scale, how long is this pathway in the actual park?" Matthew and Amayah look at the scale carefully, then calculate the actual length using the scale factor. "It's 250 feet long," Amayah says.
>
> Next, Mr. Nguyen provides students with a floor plan and asks that pairs of students choose a specific feature (e.g., length of a room, width of the hallway, etc.) and calculate its actual length using the scale.
>
> Mr. Nguyen circulates around the room, listening to the students' conversations. He hears that some students are struggling to understand the concept of

scale and how to apply it. He sits down with the students and asks, "Can you explain to me how you are using scale factor to find this length?"

The students look confused, so Mr. Nguyen encourages them to persevere. "It is okay to find it challenging, and yet, I like how you keep working to solve the problem. That's showing perseverance. Let's break it down step-by-step," he assures them.

When Mr. Nguyen is confident that students are comfortable with using the scale factor to find the actual length, he moves on. He explains to the students that they will now practice reproducing a scale drawing at a different scale. He provides the class with a new scale factor and rulers and asks students to recreate the drawing using the new scale factor. He encourages the students to use rulers and to be precise in their measurements.

Some students immediately start measuring and drawing, while others hesitate. Sitting with one group, he asks, "Could you explain how you're using the scale factor to redraw the floor plan?"

With some guidance, the students begin to understand. They erased their initial attempts and started again, this time applying the scale factor correctly. "Look, now it's beginning to look like the original drawing!" exclaims Sarah.

As the class ends, Mr. Nguyen calls the students back together. He asks some students to share their solutions and strategies with the class.

Valerie says, "I looked at the scale, which is 1 inch equals 4 feet. Then, I measured the length of the bedroom on the floor plan. It was 3 inches, so I know that the length of the bedroom is 12 feet."

Heather replies, "I changed the scale factor to 1 inch equals 6 feet, so that meant I needed to draw the bedroom length 2 inches long."

Mr. Nguyen then asks his students to talk about any challenges they faced and how they overcame them. Amayah says, "My group had trouble with the

(Continued)

(Continued)

> scale factor and understanding how to draw the floor plan. So, we looked at the original park drawing first. We saw that 1 inch was equal to 50 feet and multiplied to find that 5 inches would be 250 feet."
>
> Mr. Nguyen concludes the lesson saying, "Today, we learned how to solve problems involving scale drawings. Remember, it is important to make sense of the problem and persevere in finding a solution, just like we did today." ●

Mr. Nguyen introduced scale drawings through real-life examples like maps and floor plans, allowing students to calculate actual lengths and recreate drawings at different scales. This hands-on approach engaged students in problem-solving as they computed and reproduced scale drawings.
Mr. Nguyen fostered self-efficacy by encouraging students to believe in their ability to apply the concept and persevere as they tackled challenges with the scale factor. By working in pairs, students practiced social awareness, adjusting their approaches and sharing strategies. This vignette highlights the importance of integrating self-efficacy, perseverance, and social awareness into mathematics lessons.

Reflection

To further enrich and reinforce the lesson, incorporate reflection to deepen learning. Reflective practice is crucial for the effectiveness of any lesson, as it aids in the growth and enhancement of both the teacher and the learner. It involves the teacher and students evaluating the lesson's effectiveness and identifying areas for future improvement. This practice naturally fosters self-awareness and the synthesis of learning, engaging in higher-order thinking skills.

When reviewing the lesson, it's important to assess students' grasp of their mathematics competencies and understanding of intrapersonal and interpersonal skills. Engaging in a class discussion can be highly beneficial. There's always room for improvement, and we hope to identify and discuss positive aspects while also considering ways to enhance the learning experience.

SUMMARY

In this chapter, we examined how we can leverage MP1 and draw on the social-emotional competencies of **self-efficacy**, **perseverance**, and

social awareness to enhance classroom lessons. MP1 emphasizes making sense of problems and persevering in them. In both classroom narratives, the teachers deliberately plan and prepare to integrate content standards, practice standards, and intrapersonal and interpersonal learning competencies. They make thoughtful and purposeful pedagogical choices, employing a variety of instructional tools and methods. They highlight the integration and value of intrapersonal and interpersonal skills in their preparation. They also identify and create suitable opportunities to support student interaction, making these approaches explicit for students.

The practice of "naming and framing" intrapersonal and interpersonal skills significantly emphasizes their interplay and importance. When consistently integrated into mathematics lessons, this approach becomes a natural part of the learning process, reinforcing concepts and encouraging personal and social reflection without adding burden to the teacher or the student. Some students may have fears or beliefs about their mathematics abilities. By highlighting other skills, such as communication, adaptability, and perseverance, within the lesson, you can support students who may not feel as confident in mathematics by showing them they can use these skills to become resilient and successful.

Questions to Think About

1. How can you, as a teacher of mathematics, integrate social-emotional competencies into your lesson planning and classroom routines to support students in making sense of problems and persevering in solving them?

2. Consider a specific mathematics lesson that incorporates MP1. Which intrapersonal or interpersonal skill would most enhance the lesson?

3. How do you ensure active integration of MP1 and social-emotional competencies into your mathematics lessons?

4. Reflect on a recent mathematics lesson you taught. How could you have integrated self-efficacy, perseverance, and social awareness during the preparation and planning to positively impact your teaching and student learning?

5. What are other intrapersonal and interpersonal skills you could incorporate in this mathematics practice? What are other ways we can naturally amplify social-emotional competencies within the lesson?

Actions to Take

1. Be intentional about reflecting with your students about the skills they developed throughout the lesson. Pose the following questions to guide the discussion:
    - What mathematics skills did we develop today?
    - What other skills did you/we use to practice or learn this concept?
    - What is the value of learning this concept individually/together?
    - What social-emotional competencies did you apply in this lesson to strengthen students' understanding?

2. Reflect on how you implemented MP1 in the classroom.
    - What strategies did students use?
    - What challenges did students face?
    - How did students persevere through those challenges?

3. Discuss with your colleagues how you merge content, practices, and social-emotional competencies in your mathematics classroom, specifically related to MP1.

CHAPTER 2

FOSTERING SELF-REGULATION AND SUSTAINED ATTENTION WHILE REASONING ABSTRACTLY AND QUANTITATIVELY

MATHEMATICS IS THE ABSTRACT STUDY OF NUMBER, quantity, and space, exploring patterns and relationships through logical and quantitative reasoning. When people face real-world mathematical situations, they use quantitative reasoning to understand and solve problems, even without formal equations. School mathematics provides an opportunity for students to link real-world problem-solving with abstract reasoning using numbers, symbols, and equations.

> **REASON ABSTRACTLY AND QUANTITATIVELY**
>
> *Mathematically proficient students make sense of quantities and their relationships in problem situations. They bring two complementary abilities to bear on problems involving quantitative relationships: the ability to decontextualize—to abstract a given situation and represent it symbolically and manipulate the representing symbols as if they have a life of their own, without necessarily attending to their referents—and the ability to contextualize, to pause as needed during the manipulation process in order to probe into the referents for the symbols involved. Quantitative reasoning entails*
>
> *(Continued)*

> (Continued)
>
> *habits of creating a coherent representation of the problem at hand; considering the units involved; attending to the meaning of quantities, not just how to compute them; and knowing and flexibly using different properties of operations and objects. (CCSSM, 2010)*

Words like *abstract* and *quantitatively* can be hard to translate into lesson plans and to communicate with students, especially in elementary school. Simply put, reasoning abstractly and quantitatively means that students should be able to (1) turn a problem into a number sentence or equation (decontextualize), (2) solve it, and (3) explain the solution in the context of the problem (contextualize) as described in Table 2.1. Teachers often use student-friendly "I can" statements like, "I can connect real-world problems with numbers or equations." However, simplifying the language should not downplay the complexity of this practice, as students must integrate various types of knowledge while reasoning abstractly and quantitatively.

Table 2.1 • *Simultaneous Processes: Reasoning Abstractly and Reasoning Quantitatively*

Reason Abstractly	Reason Quantitatively
• Identify patterns and recognize mathematical structures in problem situations	• Recognize quantities or numbers
• Understand multiple ways to represent numbers	• Understand the relationship between and among numbers
• Translate the problem into a number sentence or equation	• Recognize and apply properties of operations (i.e., commutative, associative, and distributive properties)
• Recognize and understand how each number or symbol relates to the context	• Demonstrate flexibility in the use of operations and procedures
• Translate the solution back into context to make sense of the problem situation	• Not only perform mathematical calculations, but also understand why certain operations work the way they do

Mathematical Practice 2 (MP2) embeds several National Council of Teachers of Mathematics (NCTM, 2000) process standards: problem-solving, representation, and communication. First, MP2 focuses on students tackling real-world problems, developing abstract and quantitative reasoning through problem-solving (Koestler et al., 2013). Second, students must "abstract a situation and represent it symbolically" (CCSSM, 2010), highlighting how representation involves both the process and the product of capturing a mathematical concept (NCTM, 2000). Problem-solving and representation are intertwined as students decontextualize and contextualize, moving between the problem and its mathematical representation. Last, MP2 enhances mathematical communication, helping students process, organize, and clarify their thinking while gaining insights from others' reasoning and strategies (NCTM, 2000).

As students navigate between the mathematical context and the abstract representation of it, MP2 provides a chance to develop intrapersonal communication, a form of communication often undervalued and overlooked. **Intrapersonal communication**, also called self-talk, helps a person recognize emotions, gather thoughts, and guide actions. How we talk to ourselves, positively or negatively, impacts our mindset and, ultimately, our success. Intrapersonal communication also supports **self-regulation**. For example, when students encounter a challenging mathematics problem, they must regulate their emotions and manage frustration (Riegel, 2021). A student's ability to regulate themselves throughout the problem-solving process further lends itself to **sustained attention**. As students move back and forth between the problem situation and the mathematical representation, you will see the interpersonal skill of **adaptability**. Students adjust to new conditions, changes, or challenges with ease and flexibility.

MERGING CONTENT STANDARDS, MATHEMATICAL PRACTICES, AND SOCIAL-EMOTIONAL COMPETENCIES

When planning to integrate MP2 into the mathematics classroom, consider the decision points teachers encounter when selecting the lesson's content standards and goals. Identify the natural connections of MP2 to intrapersonal and interpersonal skills. Revisit the previous framework to shift the perspective toward social-emotional competencies, viewing them not as an additional task but as a fundamental element for achieving deep and meaningful learning in mathematics. This integrated approach ensures that students not only reason abstractly and quantitatively but also build essential life skills through their learning experiences.

Mathematical Content Standard and Corresponding Mathematics Goal

To illustrate the process of using an elementary content standard to answer the question, **"What is the mathematics goal of the lesson?"** consider this second-grade standard in which students "[u]se addition and subtraction within 100 to solve one- and two-step word problems involving situations of adding to, taking from, putting together, taking apart, and comparing, with unknowns in all positions, e.g., by using drawings and equations with a symbol for the unknown number to represent the problem" (2.OA.A.1, CCSSM, 2010).

In this standard, students solve addition and subtraction problems within 100 by using models like objects, drawings, bar models, number lines, or equations. They apply strategies such as counting on/back, using partial sums, or compensation. A lesson goal is for students to choose model to represent and solve a word problem and then verify that their solution is contextually accurate.

In the sixth-grade standard, students "solve real-world and mathematical problems by writing and solving equations of the form $x + p = q$ and $px = q$ for cases in which p, q, and x are all nonnegative rational numbers" (6.EE.B.7, CCSSM, 2010). The goal should be for students to identify a problem that can be represented by an equation, write and solve it, and explain their solution and approach.

Mathematical Practice

Next ask, **"Which mathematical practice supports engagement in this content standard?"** To unpack the mathematics in a standard, reexamine its language to see how it aligns with mathematical practices. Ask, "Which practice supports this content standard?" Many standards emphasize solving word or real-world problems, translating them into equations, and interpreting solutions. Terms such as *situations*, *word problems*, *story problems*, and *represent the situation* guide students in making sense of problems (contextualizing), applying necessary mathematics, representing the problem mathematically (decontextualizing), and verifying their solutions (recontextualizing). This process aligns well with MP2. Examples of other standards that align to MP2 include the following:

- Solve multi-step word problems posed with whole numbers and having whole number answers using the four operations, including problems in which remainders must be interpreted. Represent these problems using equations with a letter standing for the unknown quantity. Assess the reasonableness of answers using mental computation and estimation strategies including rounding. (4.OA.A.3, CCSSM, 2010)

- Relate the domain of a function to its graph and, where applicable, to the quantitative relationship it describes. For example, if the function h(n) gives the number of person-hours it takes to assemble n engines in a factory, then the positive integers would be an appropriate domain for the function. (HSF-IF.B.5, CCSSM, 2010)

At the elementary level, students build a foundation for reasoning abstractly and quantitatively. For example, the second-grade standard emphasizes addition and subtraction. Research shows that first and second graders grasp these concepts better using concrete objects and drawings (Bermejo & Diaz, 2007). Students can decontextualize problems using objects, then manipulate them to find solutions. As illustrated in Table 2.2, the placement of the unknown in addition or subtraction problems also affects difficulty (Bermejo & Diaz, 2007). Models help students understand solutions and how the unknown's location varies with the problem's context.

Table 2.2 • *Common Addition and Subtraction Situations*

	Result Unknown	Change Unknown	Start Unknown
Add to	Two bunnies sat on the grass. Three more bunnies hopped there. How many bunnies are on the grass now? $2 + 3 = ?$	Two bunnies were sitting on the grass. Some more bunnies hopped there. Then there were five bunnies. How many bunnies hopped over to the first two? $2 + ? = 5$	Some bunnies were sitting on the grass. Three more bunnies hopped there. Then there were five bunnies. How many bunnies were on the grass before? $? + 3 = 5$
Take from	Five apples were on the table. I ate two apples. How many apples are on the table now? $5 - 2 = ?$	Five apples were on the table. I ate some apples. Then there were three apples. How many apples did I eat? $5 - ? = 3$	Some apples were on the table. I ate two apples. Then there were three apples. How many apples were on the table before? $? - 2 = 3$

(Continued)

(Continued)

	Total Unknown	Added Unknown	Both Addends Unknown
Put together/ Take Apart	Three red apples and two green apples are on the table. How many apples are on the table? $3 + 2 = ?$	Five apples are on the table. Three are red and the rest are green. How many apples are green? $3 + ? = 5, 5 - 3 = ?$	Grandma has five flowers. How many can she put in her red vase and how many in her blue vase? $5 = 0 + 5, 5 = 5 + 0$ $5 = 1 + 4, 5 = 4 + 1$ $5 = 2 + 3, 5 = 3 + 2$
	Difference Unknown	**Bigger Unknown**	**Smaller Unknown**
Compare	("How many more?" version): Lucy has two apples. Julie has five apples. How many more apples does Julie have than Lucy? ("How many fewer?" version): Lucy has two apples. Julie has five apples. How many fewer apples does Lucy have than Julie? $2 + ? = 5, 5 - 2 = ?$	(Version with "more"): Julie has three more apples than Lucy. Lucy has two apples. How many apples does Julie have? (Version with "fewer"): Lucy has 3 fewer apples than Julie. Lucy has two apples. How many apples does Julie have? $2 + 3 = ?, 3 + 2 = ?$	(Version with "more"): Julie has three more apples than Lucy. Julie has five apples. How many apples does Lucy have? (Version with "fewer"): Lucy has 3 fewer apples than Julie. Julie has five apples. How many apples does Lucy have? $5 - 3 = ?, ? + 3 = 5$

Source: Common Core State Standards Initiative. (2010, p. 88). https://learning.ccsso.org/wp-content/uploads/2022/11/ADA-Compliant-Math-Standards.pdf

 Download this table at https://companion.corwin.com/courses/wellroundedmathstudent

Elementary students refine and expand their understanding of addition/subtraction and multiplication/division, which they then build on in middle school. This understanding extends to the algebraic manipulations students do when they "solve real-world and mathematical problems by writing and solving equations of the form $x + p = q$ and $px = q$ for cases in which *p*, *q*, and *x* are all nonnegative rational numbers" (6.EE.B.7, CCSSM, 2010). This standard is grounded in common multiplication and division situations where the location of the unknown can vary. Students can tackle many word problems using mathematical thinking and reasoning, often without relying on equations. You can harness student thinking to enhance conceptual understanding of equations by connecting students' prior experience with these problem types to build an understanding of equations in the form of $p + x = q$ and $px = q$ by revisiting common addition, subtraction, multiplication, and division situations (Tables 2.2 and 2.3).

Table 2.3 • *Common Multiplication and Division Situations*

	Unknown Product	Group Size Unknown ("How many in each group?" Division)	Number of Groups Unknown ("How many groups?" Division)
Example	$3 \times 6 = ?$	$3 \times ? = 18$, and $18 \div 3 = ?$	$? \times 6 = 18$, and $18 \div 6 = ?$
Equal Groups	There are 3 bags with 6 plums in each bag. How many plums are there in all? *Measurement example*: You need 3 lengths of string, each 6 inches long. How much string will you need altogether?	If 18 plums are shared equally into 3 bags, then how many plums will be in each bag? *Measurement example*: You have 18 inches of string, which you will cut into 3 equal pieces. How long will each piece of string be?	If 18 plums are to be packed 6 to a bag, then how many bags are needed? *Measurement example*: You have 18 inches of string, which you will cut into pieces that are 6 inches long. How many pieces of string will you have?
Arrays, Area	There are 3 rows of apples with 6 apples in each row. How many apples are there? *Area example*: What is the area of a 3 cm by 6 cm rectangle?	If 18 apples are arranged into 3 equal rows, how many apples will be in each row? *Area example*: A rectangle has area 18 square centimeters. If one side is 3 cm long, how long is a side next to it?	If 18 apples are arranged into equal rows of 6 apples, how many rows will there be? *Area example*: A rectangle has area 18 square centimeters. If one side is 6 cm long, how long is a side next to it?

(Continued)

(Continued)

	Unknown Product	Group Size Unknown ("How many in each group?" Division)	Number of Groups Unknown ("How many groups?" Division)
Compare	A blue hat costs $6. A red hat costs 3 times as much as the blue hat. How much does the red hat cost? *Measurement example:* A rubber band is 6 cm long. How long will the rubber band be when it is stretched to be 3 times as long?	A red hat costs $18 and that is 3 times as much as a blue hat costs. How much does a blue hat cost? *Measurement example:* A rubber band is stretched to be 18 cm long and that is 3 times as long as it was at first. How long was the rubber band at first?	A red hat costs $18 and a blue hat costs $6. How many times as much does the red hat cost as the blue hat? *Measurement example:* A rubber band was 6 cm long at first. Now it is stretched to be 18 cm long. How many times as long is the rubber band now as it was at first?
General	$a \times b = ?$	$a \times ? = p$, and $p \div a = ?$	$? \times b = p$, and $p \div b = ?$

Source: Common Core State Standards Initiative. (2010, p. 89). https://learning.ccsso.org/wp-content/uploads/2022/11/ADA-Compliant-Math-Standards.pdf

 Download this figure at https://companion.corwin.com/courses/wellroundedmathstudent

When connecting mathematics situations to the opportunities to reason abstractly and quantitively, you are promoting deeper understanding of the connection between mathematics and real-world situations. These connections, sense-making, and critical thinking of the real-world support social-emotional competency development.

Social-Emotional Competencies

Based on what it means to be engaged in MP2, ask, **"What intrapersonal and interpersonal skills are inherent, are needed, and can be further developed, while students engage in MP2?"** Teachers can demonstrate how to effectively incorporate intrapersonal and interpersonal skills into mathematical practice, particularly in how students communicate, understand problems, and maintain focus while solving them. Integrating these skills requires mindfulness and intentionality, ensuring they are viewed as integral elements. Therefore, it is essential to explicitly teach, scaffold, and incorporate these skills. For MP2, this includes the role of **communication, self-regulation, sustained attention,** and **adaptability** as students reason abstractly and quantitatively.

Intrapersonal Skills

MP2 naturally presents an opportunity for students to develop and strengthen **intrapersonal communication**, or inner dialogue, also called self-talk or private speech (Diaz & Berk, 1992). Self-talk serves different functions, including self-criticism (i.e., relating to negative events), self-reinforcement (i.e., relating to positive events), self-management (i.e., determining what one needs to do), and social assessment (i.e., referring to past, present, or future social interactions) (Oles et al., 2020). How we "talk" to ourselves, positively or negatively, is our constant communication or narrative within ourselves, and this "voice" has impact. Self-talk can be cyclical, impacting thoughts, behaviors, concentration, and attention (Tod et al., 2011), which also influences and impacts attitude and **self-regulation** when tackling real-world mathematics problems. For instance, students must manage frustration with challenging tasks and regulate impulses when considering others' perspectives (Riegel, 2021). This self-regulation supports **sustained attention**, allowing students to stay focused as they understand the problem, decontextualize it into an equation, perform calculations, and recontextualize the solution. This process demands time, effort, and focus.

Interpersonal Skills

Adaptability is an interpersonal skill that can be highlighted. As students move back and forth between real-world problems and abstract thinking and reasoning, this cognitive flexibility requires sustained concentration and focus. The ability to adjust strategies, reconsider approaches, and persist through challenges strengthens adaptability in mathematical problem-solving. Additionally, adaptability plays a key role in collaborative learning, as students must adjust their thinking when engaging in discussions, considering peer perspectives, and refining their reasoning. It is crucial to identify and explicitly teach one or two of these skills, so students can recognize their use and continue to strengthen them.

Instructional Structures and Engagement Strategies

To move into planning, ask, **"With an eye on our mathematics goal, how will I support social-emotional development as I engage learners** in MP2: Reason abstractly and quantitatively? What structures, strategies, methods, and/or tools can I use?"

Using Self-Talk Strategies

Before students talk about numbers, it is essential to help them develop positive intrapersonal communication, which includes self-talk and reflective thinking. These elements are critical aspects of MP2. As students contextualize a situation, they engage in self-talk to explore the relationship between quantities. Routines support self-talk in MP2 by prompting students to evaluate their initial guesses, think critically, and maintain focus. This routine encourages students to refine their estimates by considering multiple influencing factors, promoting reflection and flexibility in their problem-solving process. Through positive self-talk, students enhance their ability to engage deeply with mathematical activities.

> **Competency Builder 2.1**
> *Finding the Flipside*
>
> Children's literature provides stories that capture students' interest and attention and make abstract ideas more relatable. Read aloud *Finding the Flipside* by Jennifer Law. It is a story of a boy who thinks he cannot do anything right. Every struggle he faces leaves him feeling defeated. He is ready to quit when his spatula, Turner, encourages him to adjust his mindset. The book follows the boy through a series of struggles, including when he cannot solve a single mathematics problem and throws his pencil down. Turner tells the boy to "flip his thoughts like he flips a pancake" to see what is positive in the situation rather than focusing on the negative. Abstract thinking extends beyond the mathematics classroom, and this book effectively links the mathematical concept of the inverse with the ability to grasp it conceptually, even without direct visual representation. There are many examples of inverses or opposites in mathematics (e.g., integers, inverse operations, inverse functions, etc.). The book acknowledges the intentional effort it takes to embrace and grow optimism, persistence, and a positive mindset.
>
> After reading the book, discuss the message embedded in the story. Then, engage students in two activities:
>
> 1. Have students develop a mantra. A mantra, or a positive repeated phrase, can be said and used as a personal reminder to stay positive, focused,

CHAPTER 2 • FOSTERING SELF-REGULATION AND SUSTAINED ATTENTION

regulated, and committed to the task at hand. Have them write it down and ask them to repeat their mantra when they find themselves struggling.

2. Have students write down five things that they would tell a friend who feels challenged by the work that would help the friend stay positive and try their best. Once the students' personal lists have been generated, ask the students to read those words back to themselves before starting a hard lesson or taking a quiz or test. By "being a friend" to themselves, they can encourage themselves and promote healthier self-talk and self-efficacy.

Competency Builder 2.2
Estimation Exploration

In this routine, students view a picture and make three estimates: one too low, one too high, and one they think is accurate. This develops number sense, estimation, and the ability to contextualize and decontextualize. For example, when estimating the number of sugar cubes in a bowl (Figure 2.1), the teacher asks students for estimates and encourages reasoning with questions like, "What might be hidden?"

Figure 2.1 • *Bowl of Sugar Cubes*

Source: istock.com/MIrnaPh

(Continued)

(Continued)

This process promotes self-talk as students consider what they can't see and how that affects their estimates. They discuss their guesses with a partner, practicing communication and self-regulation by listening and considering other perspectives without immediately changing their own estimates.

To support sustained attention, instruct students to listen carefully and wait until the timer ends before revising estimates. Celebrate all reasonable estimates, not just the closest ones. To conclude, highlight the social-emotional skills displayed and summarize the key mathematical ideas, such as reasoning quantitatively and valuing different perspectives.

Some Sources for Estimation Examples

Esti-mysteries (https://stevewyborney.com/)

Estimation 180 (https://estimation180.com/)

Illustrative Mathematics: Grades K–5 Estimation Exploration (https://hub.illustrativemathematics.org/s/k5/k5-routines-list)

Using Sense-Making Activities

When students read a mathematical problem for understanding and take time to intentionally process it, this boosts learning and language acquisition (Zwiers et al., 2017). Beginning with an understanding of the situation spotlights the process of thinking abstractly and quantitatively while also amplifying learning. We can strategically teach self-regulation skills, foster sustained attention, and promote adaptability in students. By encouraging them to internally process the information they have and identify what they need to solve a problem, we help them stay focused. This focus is maintained as they seek out and apply the necessary information to reach a solution.

Competency Builder 2.3
Three Reads Routine

The Three Reads routine encourages deep engagement with word problems, prompts a reflective problem-solving process, and promotes sustained attention. Ask students how they use self-regulation skills when solving mathematics problems. Explain to students that they will process a word problem in stages. Tell them in this activity they will learn to monitor their understanding and adjust their approach at each step, helping them build self-regulation skills. As they engage in this activity, shift students' focus on different aspects of the problem during each read, demanding ongoing attention and promoting comprehension while preventing impulsive problem-solving and promoting sustained attention.

Figure 2.2 • *Three Reads Routine*

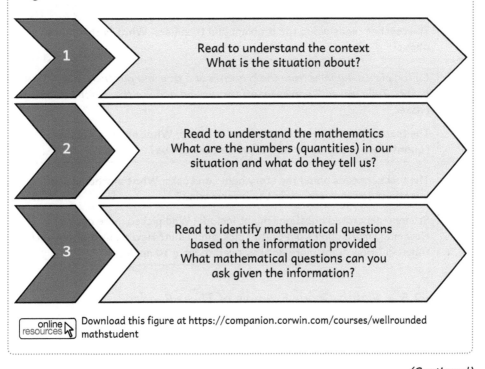

online resources: Download this figure at https://companion.corwin.com/courses/wellroundedmathstudent

(Continued)

(Continued)

This routine starts with students reading the problem and discussing the situation. They read it again with quantities, discussing what the numbers reveal. On the third read, students identify a mathematical question based on the given quantities before revealing the actual question. This approach encourages self-regulation and sustained attention, as shown in the kindergarten (Table 2.4) and eighth-grade (Table 2.5) examples.

Table 2.4 • *Kindergarten Example of Three Reads*

Three Reads	
1	*Colton picked apples from the branches of a tree and put them in a basket. Talia picked some apples up off the ground and put them in the same basket.* The teacher reads aloud the problem and then asks: **What is this story about?**
2	*Colton picked 4 apples from the branches of a tree and put them in a basket. Talia picked 3 apples up off the ground and put them in the same basket.* The teacher reads aloud the question and asks: **What are the numbers (quantities) in our story and what do they tell us?**
3	The teacher reads aloud the story again and asks: **What mathematical questions can you ask given the information?** Student-generated questions might include: Who picked more apples? How many more apples did Colton pick than Talia? How many did they pick together? How many more do they need to have 10 apples?

Table 2.5 • *Eighth-Grade Example of Three Reads*

Three Reads	
1	*Peter waited until the last minute to read his book for his book report. He finds a book and reads a few pages on day one. He reads quite a bit more on day two. On day three he reads a little less than he did on the first day, but he finishes the book!* The teacher gives the students time to read about the situation and then asks: **What is this story/problem/situation about?**

	Three Reads
2	*Peter waited until the last minute to read his book for his book report. He finds a book and reads a few pages on day one. He reads twice the number of pages on day two. On day three he reads 8 pages less than what he read on the first day, but he finishes the 214-page book!* The teacher tells the students to read the question again and then asks: **What are the numbers (quantities) in this situation? What do they mean?**
3	The teacher tells them to read the situation one more time and this time think about: **What mathematical questions can you ask given the information?** Student-generated questions might include: How many pages did he read on day one (or day two or day three)? How many more did he read on day two compared to day three?

Ask students how this strategy helps them sustain attention and maintain or regain focus. Also discuss how they can use this approach outside of mathematics. Create your own Three Reads task by removing the numbers and questions from the word problem on the first read. Add the numbers back in for the second read. Let students generate questions. Then, present them with more to answer, such as asking them to explain their reasoning, consider different methods, or apply the concepts to new situations.

Competency Builder 2.4
Information Gap

Students will develop their adaptability skills in this routine. Ask students to share a time they had to adapt their thinking, behavior, or approach to something. Explain that during this activity, they will need to adapt their approach to solving a problem and will share their strategies afterward. One student receives a problem card, and the other gets a data card. They must share ideas and information to solve the problem, which they could not do alone (Gibbons, 2002; Zwiers et al., 2017). The student with the problem card decontextualizes the situation to identify the needed information, while the student with the data card contextualizes the data. After exchanging enough information, they solve the problem independently and then compare their strategies and solutions. See Figure 2.3 for a sample student exchange.

(Continued)

(Continued)

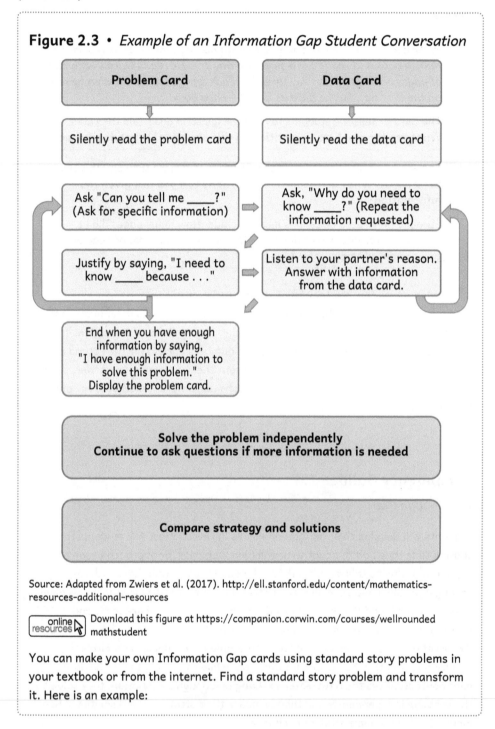

Figure 2.3 • *Example of an Information Gap Student Conversation*

Source: Adapted from Zwiers et al. (2017). http://ell.stanford.edu/content/mathematics-resources-additional-resources

Download this figure at https://companion.corwin.com/courses/wellroundedmathstudent

You can make your own Information Gap cards using standard story problems in your textbook or from the internet. Find a standard story problem and transform it. Here is an example:

> Emmalee wants to buy an off-road racing bike to compete in a local extreme sport competition. She set up a lemonade and cookie stand to raise money. Emmalee sold lemonade for $1 per cup and cookies for $2 each. She sold 60 cups of lemonade and 40 cookies. How much money did she make?

To turn this into an Information Gap, remove the quantities from the problem and use them to create the data card. What remains is a problem situation that is missing information, and this becomes the problem card (Table 2.6).

Table 2.6 • *Information Gap Problem Card and Data Card*

Problem Card	Data Card
Emmalee wants to buy an off-road racing bike to compete in a local extreme sport competition. She set up a lemonade and cookie stand to raise money. How much money did she make?	Emmalee sold lemonade for $1 per cup. Emmalee sold cookies for $2 each. She sold 60 cups of lemonade. She sold 40 cookies.

Some Sources for Information Gap Activities that Promote Adaptability

Illustrative Mathematics: Grades K–5 Information Gap
(https://hub.illustrativemathematics.org/s/k5/k5-routines-list)

Illustrative Mathematics: Grades 6–8 Information Gap
(https://hub.illustrativemathematics.org/s/68/68-routines-list)

Illustrative Mathematics: Grades 9–12 Information Gap
(https://hub.illustra tivemathematics.org/s/912/912-routines-list)

Seeing Multiple Perspectives

Teachers can use adaptability to connect multiple representations and graphic organizers in support of MP2. By promoting the use of multiple representations, teachers help students transition from concrete and pictorial examples to abstract concepts, enabling them to track their understanding and progress. Encouraging students to notice and explain relationships between different representations fosters ownership of their learning and flexibility in problem-solving strategies. This adaptability enhances their quantitative reasoning and self-regulation. Graphic organizers support MP2 by clarifying thoughts, reflecting on ideas, providing structure, and managing cognitive load. As students collaborate and encounter diverse viewpoints, these tools help them adapt their thinking and communication, fostering critical thinking and a proactive approach to learning.

> **Competency Builder 2.5**
> *Multiple Representations Rotations*

In the Multiple Representations Rotations routine, students rotate through posters in a gallery walk, analyzing and interpreting different representations of a problem using a recording sheet. Display posters like those in Figures 2.4, 2.5, and 2.6. The posters are designed to illustrate how problems can be represented in multiple ways. As students circulate and discuss the posters, they record their responses to the following prompts:

- How do the representations depict the problem?
- How do the representations connect to each other?
- What is the solution to the problem? How is the solution shown in the representations?

Figure 2.4 • *Sample Grade K–2 Multiple Representations Poster*

Five bees are in the garden. Two more are buzzing around my head. How many bees all together?

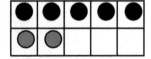

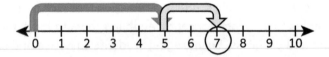

CHAPTER 2 • FOSTERING SELF-REGULATION AND SUSTAINED ATTENTION

Figure 2.5 • *Sample Grade 3–5 Multiple Representations Poster*

A recipe for a loaf of bread calls for $2\frac{1}{3}$ cups of flour. Gavin wants to make four loaves of bread. How much flour will he need?

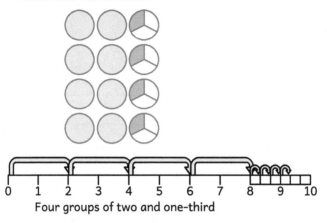

Four groups of two and one-third

Figure 2.6 • *Sample High School Multiple Representations Posters*

Andrew takes his dog to the dog park. To entertain himself, he does some counting and creates a puzzle for his mom to later solve at home: At the dog park, there were 19 dogs and people all together. I counted 62 legs. How many dogs were at the dog park?

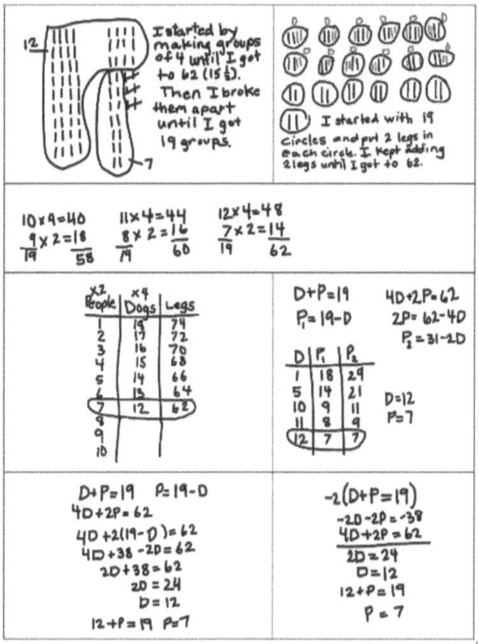

Source: Reprinted from Bay-Williams, SanGiovanni, Walters & Martinie (2022)

(Continued)

(Continued)

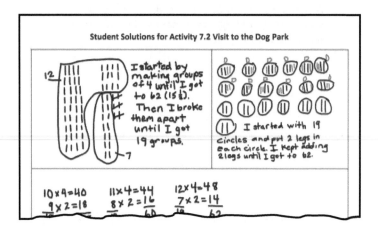

This routine leverages student discourse to connect various representations. You can support students in making connections among multiple representations to solve a contextual problem by asking probing questions as you circulate to listen and observe small groups at posters. This promotes student conceptual understanding and supports mathematical flexibility, a component of procedural fluency in mathematics.

Competency Builder 2.6
Graphic Organizers

Graphic organizers are effective tools that help students break down mathematical problems, identify missing information, and make connections. These connections enable us to adapt our approach to solving a problem. Graphic organizers help students to identify missing information and to make connections. They assist in sorting essential details, examining relationships, and organizing information for problem-solving. Graphic organizers enable students to analyze a problem's context and find a solution systematically. Figure 2.7 shows a problem-solving information organizer.

Figure 2.7 • *Problem-Solving Information Organizer*

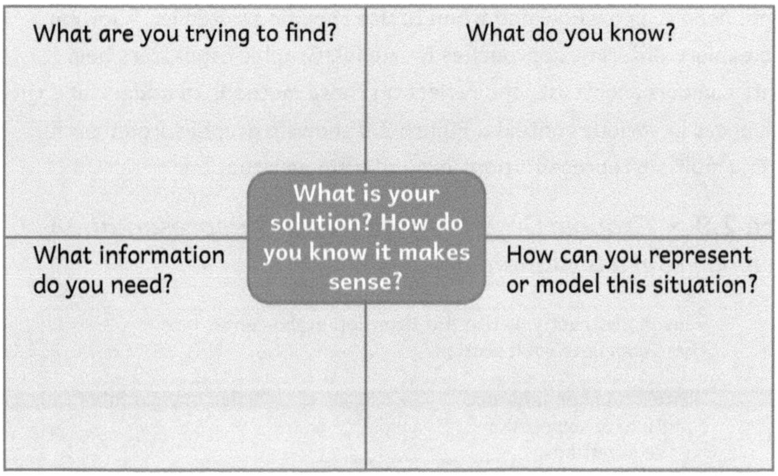

What are you trying to find?	What do you know?
What information do you need?	How can you represent or model this situation?

What is your solution? How do you know it makes sense?

Download this figure at https://companion.corwin.com/courses/wellroundedmathstudent

Graphic organizers provide a scaffolded approach to problem-solving by focusing on smaller parts, helping to clarify the connection between concrete/pictorial and abstract representations. Figure 2.8 shows a different form of graphic organizer for connecting representations to abstract reasoning.

Figure 2.8 • *Graphic Organizer for Connecting Representation to Abstract Reasoning*

Name: _____ Date: _____

CONNECTIONS IN MATHEMATICS

Directions: Draw or write your mathematical representation and number sentence in the appropriate rectangle. Then, compare and contrast the mathematical representation to the number sentence.

REPRESENTATION	NUMBER SENTENCE

Download this figure at https://companion.corwin.com/courses/wellroundedmathstudent

(Continued)

(Continued)

Students need to know how and when to use specific strategies. Allowing time to explore different approaches is crucial. Graphic organizers help students compare, contrast, and reflect on these methods to understand their effectiveness in various contexts. Figure 2.9 shows a graphic organizer for arranging multiple representations around a single situation.

Figure 2.9 • *Graphic Organizer for Multiple Representations Stemming From a Situation*

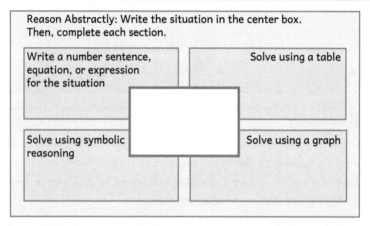

Download this figure at https://companion.corwin.com/courses/wellroundedmathstudent

Figure 2.10 shows a similar graphic organizer that connects multiple representations to a single equation.

Figure 2.10 • *Graphic Organizer for Multiple Representations Stemming From an Equation*

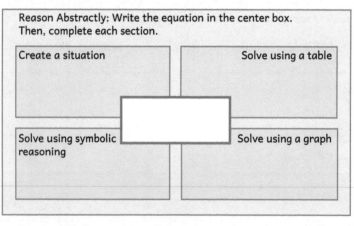

Download this figure at https://companion.corwin.com/courses/wellroundedmathstudent

These examples highlight the importance of breaking down big-picture problems into parts. Explicitly linking MP2 actions with social-emotional competencies helps students see how they support each other, fostering growth in both areas.

Assessing Interpersonal and Intrapersonal Skills

Last, ask, **"How will I assess students' progress toward the mathematics goal of this lesson, their engagement in the mathematical practice standard, and their ability to use and continue to develop social-emotional competencies? How will I provide feedback? How will I build in an opportunity for students to reflect on the development of the social-emotional competencies?"** Assessing MP2 can be challenging due to its abstract nature, but formative assessments and observations of student discussions offer immediate insights into understanding, misconceptions, and the ability to contextualize. Tools like whole class observation (Table 2.7), individual observation (Table 2.8), and self-assessment (Table 2.9) tools help track progress in mathematics skills as well as intrapersonal and interpersonal competencies.

Table 2.7 • *Whole Class Observation Tool*

Name	Mathematics Goal	Engagement in Practice Standard *Reason abstractly*	Engagement in Practice Standard *Reason quantitatively*	Intrapersonal Competency *Sustained attention*	Interpersonal Competency *Adaptability*

Note: *Progress will be marked using 0–No evidence, 1–Little evidence, 2–Adequate evidence*

Download this table at https://companion.corwin.com/courses/wellroundedmathstudent

Table 2.8 • *Individual Student Observation Tool*

Name of student:

Mathematics Goal	No Evidence	Little Evidence	Adequate Evidence
Engagement in the Practice	**No Evidence**	**Little Evidence**	**Adequate Evidence**
Reason abstractly			
Reason quantitatively			
Social-Emotional Competencies			
Sustained attention			
Adaptability			

Download this table at https://companion.corwin.com/courses/wellroundedmathstudent

Table 2.9 • *Self-Assessment Checklist*

Social-Emotional Competency	Not Sure	Not Yet	Getting There	Got It
Self-regulation				
Sustained attention				
Adaptability				
Other skills used:				

Download this table at https://companion.corwin.com/courses/wellroundedmathstudent

Observation tools effectively monitor skill development and provide feedback, but self-reflection prompts are a powerful assessment strategy to help students build self-efficacy in social-emotional competencies by encouraging regular reflection on emotions and actions. This practice fosters emotional regulation and improves behavior across various settings. Teachers should emphasize student reflection to support the development of intrapersonal and interpersonal skills. The following prompts may help students reflect on their intrapersonal and interpersonal skills:

- How did you exhibit **sustained attention** today? Provide at least one example.
- How did you practice **self-regulation** in today's mathematics activities? Provide at least one example.
- Share two examples of your **self-talk** today. Were these interactions positive or negative? If they were positive, what made them positive? If they were negative, how could you reframe your thinking to make them positive?

Alternately, teachers can provide a checklist for students to indicate their current acquisition of social-emotional competencies.

LOOKING AT EXEMPLARS IN ACTION

Now that we have explored merging content standards, mathematical practice standards, and social-emotional competencies, look more closely at an elementary and a secondary example, using the standards and competencies focused on in this chapter.

Mr. Kenyon and His Second-Grade Class

As you read the second-grade classroom narrative, note how the teacher engages students in MP2, fostering abstract and quantitative reasoning, sustained attention, self-regulation, and communication skills. Consider how you can apply these strategies in your own classroom.

The teacher, Mr. Kenyon, launches the task using the Three Reads routine, engaging students in sense-making of the mathematical task. He displays and reads aloud the following task:

> **The Splash Park Task**
>
> At the splash park, children were playing in the sprinklers. Then a group of children joined them. Now, there are more children playing in the sprinklers at the splash park.
>
> Adapted from The Swings Task (NCTM, Taking Action, K–5)

(Continued)

(Continued)

Then he asks, "What's this situation about?" Students describe it in their own words, then briefly discuss it in pairs, relating it to their splash park experiences.

Mr. Kenyon brings them back together and rereads the problem with numbers included.

> **The Splash Park Task**
>
> At the splash park, 15 children were playing in the sprinklers. A group of children joined them. Now, there are 21 children playing in the sprinklers.
>
> Adapted from The Swings Task (NCTM, Taking Action, K–5)

He displays these questions on the board: "What are the quantities (numbers) in our story? What do they mean? How are they related?" He asks students to work in pairs and share their ideas. They identify 15 as "children playing in the sprinklers" and 21 as "children playing in the sprinklers," explaining that 15 were there at the beginning and then 21 because more joined. They discuss the relationship using phrases like *starting with*, *added to*, and *total*.

Mr. Kenyon regroups students, and everyone reads the problem aloud again. Mr. Kenyon asks, "What mathematical questions can we ask?" Students discuss with a partner to identify possible questions.

At this point, he displays the complete task, revealing the questions, "How many children joined the first group? How can this problem be drawn or represented? Be prepared to explain your solution."

> **The Splash Park Task**
>
> At the splash park, 15 children were playing in the sprinklers. A group of children joined them. Now, there are 21 children playing in the sprinklers. How many children joined the first group?

> How can this problem be drawn or represented? Be prepared to explain your solution.
>
> Adapted from The Swings Task (NCTM, Taking Action, K–5)

The following presents two excerpts from a larger class narrative illustrating how the teacher supported students' engagement in MP2 and social-emotional competencies.

Liam and Zellie are working together when Liam gets distracted. Zellie refocuses him, saying, "Liam, we have to do this!"

Liam asks about the number of kids who are playing in the sprinkler. He asks Mr. Kenyon, "Can 21 kids even play in the sprinkler at the same time?"

Mr. Kenyon laughs and asks, "What are you thinking?"

Liam imagines their splash park and wonders if there is enough space for all those children. Mr. Kenyon acknowledges his thought and redirects him, suggesting they figure that out at another time. For now, they should imagine a bigger splash park.

Liam refocuses and says, "I think we can find how many kids play in the sprinklers all together."

Zellie disagrees, and Mr. Kenyon prompts her to use a sentence starter. She explains, "I disagree because we already know how many are playing in the sprinklers. We need to find how many children join them." She identifies it as a "put together" situation and writes the equation (___ + ? = ___).

Liam and Zellie then share their plans and begin solving independently. ●

While using the Three Reads routine, Liam and Zellie demonstrate **sustained attention**. Liam briefly strays, connecting to his own experiences as he imagines his community's splash park. He then **self-regulates** and refocuses on the task. Zellie reframes her initial reaction using a language frame, respectfully **communicating** her understanding of the addition problem.

Instead of giving direct instructions, Mr. Kenyon acknowledges Liam's curiosity. He also prompts Zellie to use language frames, reinforcing structured conversation. This practice not only deepens their understanding of addition but also prepares them to apply these skills in other contexts.

As Mr. Kenyon circulates the classroom, he observes students' work and listens to their discussions on addition and subtraction. He notices Maya has incorrectly written "$15 + 21 = 36$" on her paper. When he asks her to explain how this number sentence relates to the situation, she realizes her mistake after rereading the problem.

Maya corrects herself, writing "$15 + ? = 21$." Maya begins at 15, writing and counting aloud as she records the numbers up to 21, but she miscounts and arrives at 5 instead of 6. Mr. Kenyon prompts her to prove her solution using another model. By using a number path, Maya counts all the students, leading her to revise the answer to 6. She confirms this by recounting.

Mr. Kenyon asks Maya to summarize her thinking, and she explains how she overcame her frustration by using calm breathing and positive self-talk. He commends her for her determination and focus in solving the problem.

When Maya used a model to prove or disprove her equation, she exhibited **sustained focus**. In the face of an incorrect solution, she displayed **self-regulation** by checking her emotions and controlling her behavioral response. When an incorrect solution discouraged Maya for a second time, she accessed a strategy to calm her body and mind. Maya's **intrapersonal communication** supported her sense-making and self-regulation.

Mr. Kenyon could have made a mental note that Maya seemed to self-regulate herself very well. Instead, he seized a moment to engage Maya in a conversation that applauded her access to a **self-regulation** strategy. This conversation encourages Maya to access this strategy again. He also emphasized her **sustained attention**, complimenting her focus, attention, and determination while also praising her positive **intrapersonal communication**.

Mr. Thompson and His Sixth-Grade Class

While reading the sixth-grade classroom narrative, note how the teacher engages students in MP2 and helps students develop sustained attention, self-regulation, and communication skills. Consider how you apply these strategies in your own classroom.

In Mr. Thompson's sixth-grade math class, students are learning to write and solve equations. To prevent impulsive guessing, he uses the Three Reads routine (Kelemanik et al., 2016). He displays and reads aloud the following task:

> **Saving up for Sports**
>
> Kalob wants to save money to buy new sports equipment. He found a set that includes a football, basketball, baseball, and soccer ball. To earn money, he plans to wash neighbors' windows.

Mr. Thompson asks, "What is this situation about?" and has students describe it in their own words. Students briefly discuss in pairs, relating the context to their own experiences saving money.

Then, Mr. Thompson brings them back together and rereads the problem, now including numbers.

> **Saving up for Sports**
>
> Kalob wants to save money to buy new sports equipment. He found a set that includes a football, basketball, baseball, and soccer ball for $108 after tax. To earn money, he plans to wash neighbors' windows. He will charge $4 for each window he washes.

He displays these questions on the board: "What are the quantities (numbers)? What do they mean? How are they related?" Students pair up to identify $108 as the cost of a sports equipment set and $4 as the amount earned per window washed. They discuss how these quantities are related, using terms like *$4 each*, *multiplying*, and *total*.

Mr. Thompson then rereads the problem for a third time and asks, "What mathematical questions can we ask?" Students discuss and identify, "How many windows must he wash to save enough for the sports equipment set?"

Finally, he reveals the full task and asks students to solve it independently, write an equation, and explain how it represents the situation.

(Continued)

(Continued)

> ### Saving up for Sports
>
> Kalob wants to save money to buy new sports equipment. He found a set that includes a football, basketball, baseball, and soccer ball for $108 after tax. To earn money to purchase the sports equipment set, he plans to wash neighbors' windows. He will charge $4 for each window he washes. How many windows must he wash to save enough for the sports equipment set?
>
> - Solve the problem in any way that makes sense to you.
> - Write an equation to represent the situation.
> - Explain how the equation represents the situation.
>
> Adapted Task (Illustrative Mathematics)
>
> http://tasks.illustrativemathematics.org/content-standards/7/EE/B/4/tasks/986

As Mr. Thompson circulates the classroom and monitors student progress, he finds Braeden with some work on his paper, but it is incomplete. Braeden appears either deep in thought or daydreaming. Mr. Thompson asks, "Do you need help?"

Braeden snaps back to attention and says, "Oh, no, I don't need help. I just need to focus." Mr. Thompson prompts Braeden to say more. Braeden elaborates, "Well, I was thinking about the sports set and washing windows to earn money. That made me think about my neighborhood and washing windows. Anyways, sorry. I'll find the answer now." Mr. Thompson affirms his refocus as Braeden begins working again.

After a short time, Mr. Thompson circulates back to Braeden, checking on his progress. Mr. Thompson asks if Braeden has reached the goal. Braeden says, "Almost. I'm not done yet. I need to explain my equation, and then I will be finished." Braeden writes out the meaning for each component of his equation $4 \times w = 108$.

Another student, Kieryn, quickly starts solving by trying different operations with 4 and 108, writing 432 and $4 \times 108 = 432$. Mr. Thompson prompts her to explain her process and reminds her to think through the situation

> rather than randomly guessing operations. Kieryn admits she thought it was multiplication but realized 432 seemed high, so she tried other operations.
>
> Mr. Thompson acknowledges her number sense and advises her to pause and think when stuck. They revisit the problem, and after discussing the quantities, Kieryn divides 108 by 4 and writes the equation $108 = w \times 4$.
>
> When asked if she met the goal, Kieryn is unsure. Mr. Thompson prompts her to reread the directions and check her work. Kieryn realizes she needs to explain how the equation fits the problem. She explains that *4* represents the money per window, *108* is the cost of the sports equipment, and *w* is the number of windows. Mr. Thompson asks when she'll know she's done, and Kieryn says she'll know she's done after explaining the meaning of each part of her equation.

When Mr. Thompson asks Braeden how he can help, Braeden **self-regulated**, recognizing his brief inability to focus rather than needing the teacher's help with the mathematics at hand. When Mr. Thompson prompts Kieryn to think before trying different number and operation combinations, he encourages **self-regulation** to manage impulsivity. When he integrates the goal of the lesson, he prompts **sustained attention** and **self-regulation**. When he prompts Kieryn to share her thinking process, it reveals her intrapersonal **communication**.

Reflection

Across both classroom narratives, the teachers intentionally plan and prepare for the integration of content standards, practice standards, and intrapersonal and interpersonal learning competencies—specifically self-regulation, sustained attention, and intrapersonal communication. The teachers in these narratives do this with thoughtful and purposeful pedagogical choices while utilizing a variety of instruction tools and methods. Specifically, they provide structures to support students' abstract and quantitative reasoning. They designed the lesson to include real-world mathematics that also explicitly integrate social-emotional competencies. These moments of supporting self-regulation, sustained attention, and communication did not take an exorbitant amount of time to act. The teachers' responses amplified the opportunity to support intrapersonal and interpersonal skill development.

SUMMARY

In this chapter, we examined how teachers leverage MP2 and draw on social-emotional competencies to enhance their classroom lessons. MP2 helps students to reason abstractly and quantitatively, emphasizing the importance of making sense of a situation (contextualization) and relating the numbers to the situation (decontextualization). By connecting the context and content, students can make sense of quantities and their relationships in problem situations and the mathematics in their world. When teachers focus on reasoning abstractly and quantitatively, students naturally extend this skill beyond the classroom walls.

Aligning mathematics lessons with MP2 by using mathematical standards like 2.OA.A.1 and 6.EE.B.7 promotes students' ability to solve real-world mathematics problems while also emphasizing the process of (1) translating a problem situation into a number sentence or an equation (decontextualize), (2) performing the arithmetic or algebra, and then (3) revisiting the problem situation and explaining what the solution means (contextualize).

The integration of intrapersonal and interpersonal skills, such as intrapersonal communication, self-regulation, and sustained attention, all focus on the ability to remain focused and attentive to the task at hand. The interplay of context and content requires sustained attention, monitoring of self, and the ability to adjust to the given circumstances.

In our classroom narratives, Mr. Kenyon's class engages in a mathematics lesson about adding or subtracting to solve one- and two-step word problems, encouraging intrapersonal communication, self-regulation, and sustained attention. In Mr. Thompson's class, students write and solve equations for real-world situations, applying what they already know about solving the problem and writing equations that represent the situation, providing opportunities for intrapersonal communication, self-regulation, and sustained attention.

Questions to Think About

1. Revisit Table 2.1 from the introduction of this chapter. In what ways did these teachers address the two aspects of this practice: reason abstractly and reason quantitatively? What role did the social-emotional competencies of self-regulation, sustained attention, and intrapersonal communication play in supporting these actions?

2. Reflect on a recent mathematics lesson you taught. How could you have integrated communication, self-regulation, and sustained attention in the preparation and planning to positively impact your teaching and student learning?
3. What are other intrapersonal and interpersonal skills you could incorporate in this mathematical practice?
4. In what ways do you provide students feedback on the development of social-emotional competencies?

Actions to Take

1. Write any of these questions on a sticky note:
   - How will I support intrapersonal communication in an upcoming lesson?
   - How will I support self-regulation in an upcoming lesson?
   - How will I support sustained attention in an upcoming lesson?

   Post this sticky note wherever you write lesson plans to remind yourself to incorporate these social-emotional competencies into an upcoming lesson.

2. Select an open task from a previous grade level to activate students' prior knowledge. (Consider a previous grade-level standard that will support the new topic in your grade-level standards.) Then, decide which intrapersonal and interpersonal skills to support as students navigate the open task. Focus teacher feedback on supporting the desired intrapersonal or interpersonal skill(s). This helps students practice social-emotional skills within a familiar mathematics standard.

3. Practice aspects of these intrapersonal and interpersonal skills in nonmathematical situations. (For example, ask students to engage in conversations using language frames. Select a nonmathematical prompt, such as "What is the best television show?" to practice these sentence starters.)

4. Give a contextualized mathematics task. While students engage in abstract and quantitative reasoning, support intrapersonal and interpersonal skill development.

5. When students practice or display a social-emotional competency (such as communication, self-regulation, or sustained attention), provide specific feedback by identifying and naming the skill, describing the action, and reinforcing or redirecting the action.

CHAPTER 3

GROWING SELF-AWARENESS AND SOCIAL AWARENESS THROUGH CONSTRUCTING AND CRITIQUING ARGUMENTS

THINK ABOUT THESE WORDS: *argue, debate, fight.* While they all involve conflict, they differ in how they approach conflict. Turning a conflict into a constructive conversation allows children to build valuable skills, take responsibility for their thoughts, and own their ideas. Constructing and critiquing mathematical arguments in the classroom provides meaningful opportunities for students to engage with mathematics. Research shows that focusing on argumentation in mathematics helps students become more independent learners, boosts their confidence, helps them find personal meaning in mathematics, and increases their engagement (Boaler, 2019; Brown, 2017; Francisco & Maher, 2005; Yackel & Cobb, 1996).

Argumentation in mathematics class emphasizes student justification, which shifts attention from answer-getting to sense-making. According to Boaler (2019), "explaining work is what we call mathematical reasoning, and if students are not reasoning, they are not being mathematical." Constructing and critiquing arguments are metacognitive skills that students need support to develop. This involves clear teaching of the necessary knowledge, skills, and behaviors (both intrapersonal and interpersonal), modeling by teachers and peers (Schneider & Artelt, 2010), and providing opportunities for practice. Specifically, students need to come to understand what it means to "construct viable arguments" and to "critique the reasoning of others." While Mathematical Practice 3 (MP3) describes clear actions for constructing and critiquing mathematical arguments, these actions are transferable to many everyday situations.

CONSTRUCT VIABLE ARGUMENTS AND CRITIQUE THE REASONING OF OTHERS

Mathematically proficient students understand and use stated assumptions, definitions, and previously established results in constructing arguments. They make conjectures and build a logical progression of statements to explore the truth of their conjectures. They are able to analyze situations by breaking them into cases and can recognize and use counterexamples. They justify their conclusions, communicate them to others, and respond to the arguments of others. They reason inductively about data, making plausible arguments that take into account the context from which the data arose. Mathematically proficient students are also able to compare the effectiveness of two plausible arguments, distinguish correct logic or reasoning from that which is flawed, and—if there is a flaw in an argument— explain what it is. Elementary students can construct arguments using concrete referents, such as objects, drawings, diagrams, and actions. Such arguments can make sense and be correct, even though they are not generalized or made formal until later grades. Later, students learn to determine domains to which an argument applies. Students at all grades can listen to or read the arguments of others, decide whether they make sense, and ask useful questions to clarify or improve the arguments. (CCSSM, 2010)

How can the two aspects of this practice be explicitly addressed? Begin by explaining to students that they will use their knowledge to create and communicate their own problem-solving ideas while also listening to and evaluating others' ideas. Plan lessons that include these actions, discuss and model them, and reflect on student engagement. Identify specific actions for each aspect, show how intrapersonal and interpersonal skills are included, and share this information with students through handouts or classroom posters (see Table 3.1).

Table 3.1 • *Actions Associated With MP3*

Construct Viable Arguments	Critique the Reasoning of Others
• Know what the problem is asking based on what you are given • Consider strategies or approaches to solving the problem • Choose a strategy or approach to employ • Determine whether the results make sense • Communicate to others what you did and why it makes sense	• Actively listen to the explanation • Ask questions to better understand what they did and why they did it • Identify any errors or missing information and respond respectfully • Compare the strategy/approach to others

online resources Download this table at https://companion.corwin.com/courses/wellroundedmathstudent

As students construct their own arguments, receive feedback on their ideas, and provide feedback to others, there are many ways that you can both leverage and develop intrapersonal and interpersonal skills. Generating their own ideas and sharing them involves risk-taking for students, offering an opportunity to address **self-awareness**. Because sharing involves risk-taking, students need to learn to show concern for the feelings of others through compassion and **empathy**. To do this, students need to leverage and continue to develop **social awareness**, acknowledging and appreciating that others may have a different perspective. Additionally, students are asked to evaluate arguments and decide whether they make sense, as well as ask useful questions to clarify or improve the arguments. Developing this sort of **decision-making** requires them to **assert** their own thoughts and ideas in respectful ways while being treated with respect when doing so.

At all levels, being purposeful in planning for this mathematical practice is critical and this can be done by (1) connecting the mathematics content with the mathematical practices, (2) relating mathematics to one's own intrapersonal and interpersonal development, and (3) considering what is developmentally appropriate for the age group you are teaching. In the introduction we provide a framework for doing this.

MERGING CONTENT STANDARDS, MATHEMATICAL PRACTICES, AND SOCIAL-EMOTIONAL COMPETENCIES

Tasks that involve constructing arguments and critiquing other people's reasoning can be implemented in a way that develops mathematics concepts and at the same time develops and leverages essential interpersonal and intrapersonal skills. Even tasks that are noncurricular in nature help develop the critical thinking skills and mindset that enable students to engage more productively in subsequent curricular tasks. Consider again the framework and question prompts presented in the introductory chapter.

Mathematical Content Standard and Corresponding Mathematics Goal

Begin by asking, **"What is the mathematics goal of the lesson aligned to this standard?"** Consider a fourth-grade standard: "Decompose a fraction into a sum of fractions with the same denominator in more than one way, recording each decomposition by an equation. Justify decompositions, *e.g. by using a visual fraction model*" (4.NF.3b, CCSSM, 2010). Students investigate decomposing fractions beyond the use of just unit fractions. The goal is for them to know and understand that the numerator of the fraction indicates the number of unit fractions we make when the fraction is decomposed.

For example, $\frac{3}{4} = \frac{1}{4} + \frac{1}{4} + \frac{1}{4}$

Notice that the number of iterations corresponds to the numerator's value, illustrating the number of equal parts that make up the fraction. However, fractions can be decomposed in various ways beyond unit fractions and often can be broken down in multiple ways.

For example, $\frac{7}{10} = \frac{1}{10} + \frac{2}{10} + \frac{4}{10}$ or $\frac{7}{10} = \frac{3}{10} + \frac{4}{10}$

Moreover, students are expected to justify the decompositions by showing and explaining how different models of the fraction are equivalent.

At the high school level, consider the standard "solve quadratic equations choosing from several methods: by inspection (e.g., for $x^2 = 49$), taking square roots, completing the square, the quadratic formula and factoring, as appropriate to the initial form of the equation" (HS.REI.B.4b, CCSSM, 2010). One goal is to develop procedural fluency solving quadratic equations. This involves not only teaching students how to use a particular method, but also when it is appropriate and most useful. Fundamentally, solving quadratic equations is a reasoning process. Achieving this standard means that students understand the reasoning behind these methods, recognize the relationships among them, and can apply them in the right circumstances. For example, students should recognize that completing the square builds on a combination of factoring and taking square roots. Students need the opportunity to recognize these structures and consider the advantages and disadvantages or strengths and weaknesses of these methods.

Mathematical Practice

After examining the language of the standard to unpack the mathematics, look specifically for language that aligns to the mathematical practices. Ask, **"Which mathematical practice enhances understanding of this content standard?"** Identify the language in the standard that implies a variety of approaches or models and requires students to decide how or when to use them. Analyzing the type of thinking involved in the standards enables teachers to connect the content to a practice standard. Additional examples of such standards include the following:

- Decompose numbers less than or equal to 10 into pairs in more than one way, e.g., by using objects or drawings, and record each decomposition by a drawing or equation (e.g., $5 = 2 + 3$ and $5 = 4 + 1$). (K.OA.A.3, CCSSM, 2010)
- Decide whether two quantities are in a proportional relationship, e.g., by testing for equivalent ratios in a table or graphing on a coordinate plane and observing whether the graph is a straight line through the origin. (7.RP.A.2.A, CCSSM, 2010)

Notice these content standards require that students know how to do something in more than one way. This means they need to know how to use these strategies, methods, or models and when to choose them. This provides an opportunity to develop and use intrapersonal and interpersonal skills in the mathematics classroom as students justify their choices, make an argument to explain and justify their solution process, communicate ideas to others, and respond to the arguments of others.

Social-Emotional Competencies

Next ask, **"What intrapersonal and interpersonal skills are inherent, are needed, and can be further developed for students while engaging in MP3?"** MP3 relies heavily on communication (listening, justifying, communicating, responding, explaining). These communication skills have intrapersonal and interpersonal skills associated with them. We will focus specifically on **self-awareness**, **decision-making**, **social awareness**, **empathy**, and **assertiveness**.

Intrapersonal Skills

As students "listen to or read the arguments of others, decide whether they make sense, and ask useful questions to clarify or improve the arguments," they must engage in **self-awareness**. Self-awareness happens when students find themselves challenged at times by the feedback of their peers. It requires them to better understand and reflect on their own feelings, thoughts, values, and behaviors. They must notice if they feel defensive and let go of that defensiveness, which allows them to view themselves in new ways. Constructing an argument and critiquing reasoning is closely related to **decision-making** in that both processes are about weighing options and arriving at reasoned conclusions. They both involve clarity of thought, evidence and support, strategic thinking, reasoning, and reflection.

Interpersonal Skills

MP3 is also heavily dependent on **social awareness**—students' ability to recognize others' perspectives. As they listen to their peers' arguments, they need to demonstrate understanding and appreciation for the ideas shared and learn to consider the diverse contexts and backgrounds that influence others' perspectives. MP3 presents an opportunity for students to practice disagreeing while still respecting each other's perspective. Sharing involves risk-taking and, therefore, students need to learn to show concern for the feelings of others through compassion and **empathy**. Responding to the arguments of others requires trying to understand and showing you understand whether you agree or not. Students also evolve a sense of **assertiveness** as they state their own thoughts and ideas in respectful ways.

Instructional Structures and Engagement Strategies

As your focus moves toward planning a lesson, the question arises, **"With an eye on our mathematics goal, how will I support social-emotional development as I engage learners in MP3: Construct viable arguments and critique the reasoning of others? What structures, strategies, methods, and/or tools can I use?"**

MP3 calls on students to "communicate" their ideas with others and "respond to the arguments of others." Children learn language, in this case mathematical language, by using it and by hearing it used (Prediger et al., 2019; Schleppegrell, 2007). A well-designed task and an opportunity to work collaboratively sets the stage for students to communicate about mathematics as they grapple with it and engage in decision-making. This supports the development of the academic language necessary for them to construct viable arguments and critique the reasoning of others. It also promotes the development of social awareness and the skills of empathy and assertiveness. To develop these skills, students need to attend specifically to how to construct, justify, and critique an argument.

Defining Social Awareness

Cooperative and collaborative learning structures support engagement in collective argumentation, where the teacher and students work together to formulate ideas and provide justification (Connor et al., 2014). Students leverage and develop social awareness as they engage in MP3. However, the skills embedded in this competency need to be made explicit.

Competency Builder 3.1
Brainstorming Social Awareness

Begin by defining social awareness. Explain that social awareness involves the ability to empathize with others' experiences and the willingness to seek to understand their perspective (Greater Good Science Center, 2024). Engage students in brainstorming skills related to social awareness and organize the ideas into a table. Figure 3.1 provides a list of possible skills. The language used to express these ideas will vary depending on the age of the students.

Figure 3.1 • *Social Awareness Skills (Greater Good Science Center, 2024)*

Social Awareness
• Identify social cues to determine how others feel
• Examine the perspectives of others
• Ask questions to better understand others
• Show empathy and concern
• Express gratitude and appreciation
• Recognize and appreciate others' strengths
• Take action to contribute to the well-being of others
• Care about the greater good

Download this table at https://companion.corwin.com/courses/wellroundedmathstudent

(Continued)

(Continued)

> Reference this table frequently and model these skills. Help students identify them in action by calling them out when students are working, saying things like,
>
> - I see you put your pencil down and you look upset, are you feeling frustrated?
> - Thank you for helping your partner when he was stuck.
> - How do you feel now that you have finished this problem?
> - If someone in your group has not had a chance to share, please invite them to do so by asking them what they think.
>
> Social awareness and relationship skills go hand-in-hand. Being aware of the thoughts and feelings of others enables students to collaborate more productively with each other, communicate effectively, and resolve conflicts.

Constructing Arguments

On a cognitive and personal level, constructing an argument is challenging for students, and communicating an argument is even more challenging. This challenge is compounded by the difficulties teachers experience engaging students in constructing arguments and responding to those of others (Ayalon & Even, 2016; Ayalon & Hershkowitz, 2018). From a teaching perspective, you can look to the description of the mathematical practice standard for suggestions for addressing this challenge. For example, consider, "Elementary students can construct arguments *using concrete referents such as objects, drawings, diagrams, and actions.*" You can encourage students to leverage these tools to build understanding of mathematics concepts in and beyond elementary grades (Gulkilik, 2016; Larbi & Mavis, 2016).

> ## Competency Builder 3.2
> *Leveraging Concrete Representations*
>
> Concrete models support conceptual development, engagement in MP3, and building of social-emotional competencies across all grade levels. Begin by selecting a representation that aligns the mathematics concept you are teaching.

For example,

- In a fraction lesson, students use fraction tiles to build and decompose fractions. They then use their representation to argue about the equivalency of fractions they have constructed and critique each other's reasoning.
- Students use physical models of 2D and 3D shapes to explore their properties and use the physical models to test claims like, "All squares are rectangles, but not all rectangles are squares." This supports peer critiques that encourage respectful communication, active listening, and constructive feedback.
- Students use Algebra Tiles to represent and solve linear equations, multiply binomials, factor trinomials, or to complete the square as a technique to solve quadratic equations. Often, algebra instruction focuses heavily on algebraic manipulation, and students struggle to understand what is going on conceptually. Algebra Tiles allow students at all developmental levels to have access to important understandings and more clearly see what others are explaining.
- At all levels, students can draw diagrams to represent word problems, which allows them to both show and talk about their problem-solving process with their peers. Tools like these help students learn to explain their reasoning and consider other perspectives in a safe and supportive environment.

Engage students in a lesson using representations that align to the concepts you are addressing. As a class, model how to use the tools, how to make a claim or construct an argument, and how to critique each other's solutions. Pair or group students to work on the tasks. Reconvene the class for a discussion. Ask for volunteers to share how they used the tools to do the mathematics and to justify the steps of the process. End by having students reflect (in writing or using a Think-Pair-Share) on the activity and the group interactions. Ask questions like the following:

- Why is it important to use concrete models or diagrams like these to justify your answers? (decision-making)
- How did you feel when someone critiqued your reasoning? (self-awareness)
- How do you handle disagreements when you are solving problems with peers? (social awareness)
- How do you point out a mistake someone has made in a way that helps them learn but also not feel bad about it? (empathy)
- What strategies did you use to help your peers understand your thinking or your work? (assertiveness)

(Continued)

(Continued)

> Questions like these only take a few minutes, but the impact is beyond measure. The combination of hands-on practice, peer critique, and justification encourages deep mathematical understanding, engagement in mathematical practices, and development of social-emotional competencies.

Competency Builder 3.3
Notice and Choose Routines

A Notice and Choose routine is an extension of a Notice and Wonder routine and is used to support students in generating ideas. This routine both leverages and develops assertiveness. Consider the standard "students solving quadratic equations choosing from several methods: by inspection, taking square roots, completing the square, the quadratic formula and factoring, as appropriate to the initial form of the equation" (HS.REI.B.4b, CCSSM, 2010). This involves noticing structures of an equation that would lend itself to a particular solution method. Explain to students they will need to be assertive when making and explaining their choice. Remind them that assertiveness means they clearly state their own thoughts and ideas in respectful ways. Ask them why it is important to be assertive when making choices. Provide students with several quadratic equations that vary in form and draw on a variety of methods.

| $2x^2 + 8x + 6 = 0$ | $9x^2 + 10 = 91$ | $12(x - 4)^2 - 2 = 0$ |

Then ask, "What do you notice? How can this help you efficiently choose a method to solve the quadratic equation? How can you justify your choice?" Asking students what they know about familiar content has been shown to effectively launch students into making claims (Rumsey & Langrall, 2016). We use these routines to teach students to pause and examine or explore the task at hand before they begin working. This supports not only mathematical thinking and reasoning but also supports thinking about options before asserting ideas and making decisions in general.

To conclude the routine, prompt students to consider the actions they took during the routine to make it work effectively. There are several actions students may suggest, such as they had to be curious and/or creative, observant, consider options, discuss ideas with others, take risks to share ideas, and justify their choices. This routine creates a low-stakes environment where everyone can make claims based on their perspective. Sharing this enables others to expand ideas beyond what they think, and this promotes self-awareness and social awareness.

Justifying Arguments

Children, and especially teenagers, love to argue. But what makes good justification in an argument? As early as first grade, students are expected to "write opinion pieces in which they introduce the topic or name the book they are writing about, state an opinion, supply a reason for the opinion, and provide some sense of closure" (W.1.1, CCSS.ELA-Literacy, 2010). In middle and high school, this grows into a more formal process for making claims, supporting these claims with evidence, and drawing analytical conclusions. Helping students understand what it means to justify in mathematics begins by having them justify decisions, actions, or ideas in situations they have experienced outside of mathematics. In this way, this practice enhances the development of decision-making.

Competency Builder 3.4
What Makes Good Justification?

If you are an elementary teacher, think about what your students know from their ELA lessons about what makes for a good argument. If you are a secondary teacher, collaborate with ELA teachers to access students' prior knowledge regarding constructing arguments, and then expand on it to include mathematical claims supported with mathematical evidence resulting in a mathematical conclusion. Create a chart or template that identifies four main parts of a mathematics argument (idea, justification, feedback, and conclusion) to make these pieces explicit for students. For example, you can use the template in Figure 3.2 to include incorporating feedback and using constructive criticism when creating an argument. Before doing this in a mathematics lesson, use the template to address a noncurricular situation that students would be familiar with, such as "When is the best time for recess?" or "Should phones be allowed in class for educational purposes?" To emphasize decision-making, ask them to

(Continued)

(Continued)

reflect on the steps in the template and the importance of each step in making and justifying a claim. In earlier grades, the template can be completed in small groups while you record the ideas or as a whole class to model this type of thinking.

Figure 3.2 • *Creating an Argument Template*

Creating an Argument

Should phones be allowed in class for educational purposes?

Explain and/or show your idea.
Yes, because learning with technology can make learning fun.

Justify your idea (i.e., show or explain what makes it reasonable).
Phones are a kind of technology. They are easy to use, and everyone (almost) has one. There are ways to practice that are like a game.

Incorporate feedback. What did your peers or others suggest that you could use to improve your idea?
Just because it is fun doesn't mean you can learn from it. It can also be distracting because you can play other games that aren't for learning.

Reflect. Based on the input you have collected, what is your final idea?
Yes, because technology can help you learn better by showing examples of how to do things that are different from what the teacher shows you. There are also educational games. You can also learn to be more responsible and use your phone the way you are supposed to at school.

Source: Adapted from Knudsen et al., 2014.

 Download this figure at https://companion.corwin.com/courses/wellrounded mathstudent

To create a mathematical argument, students begin with an idea or response they generated (first box) and convince themselves their idea is correct. Provide individual think time and ask students to record what makes this reasonable (second box). Prompt students to represent their idea in multiple ways (visually, pictorially, verbally, numerically, tabularly, and algebraically) to make their reasoning visible to others. Have students present their idea and their reasoning to a peer and receive feedback. Display questions students can ask when they are presented with an argument and ask them to practice using the questions or to generate their own (e.g., "What made you choose to do that? How do you know it works? Where did you get that number? What does this represent?"). Finally, have students reflect on the feedback and make a final decision.

Conclude the lesson by explaining to students they can develop self-awareness if they stop and reflect on their own emotions, thoughts, behaviors in this lesson, and how they influenced their own actions and that of their peers. Explain that self-awareness happens when you find yourself challenged by the feedback of your peers. For example, you might notice if you feel defensive and let go of that defensiveness, which allows you to view yourself in new ways. Have students think about two different emotions they felt during the lesson and how they affected their behavior. Ask for volunteers to share. Summarize by saying that self-awareness plays an important role in personal growth, decision-making, and relationship-building, and self-reflection is a good way to improve self-awareness.

Critiquing Arguments

Critiquing arguments requires communication skills and empathy. Tools such as language frames (Figure 3.3) support the development of social awareness by equipping students with sentence starters that foster positive communication and are an effective strategy for practicing discourse focused on argumentation (Rumsey & Langrall, 2016) and promoting assertiveness.

Figure 3.3 • *Sample Language Frames*

I agree with _____ because _____.
I notice _____ when _____.
I wonder why _____.
I have a question about _____.
I disagree because _____.
Based on _____, I think _____.

In addition, critiquing arguments involves producing counterexamples for a conjecture and is delivered more effectively with empathy. A conjecture is a statement based on what one knows or observes and is often formed with incomplete information. A counterexample opposes or contradicts a conjecture. Counterexamples exist all around us in the real world and are very important when developing reasoning, formulating ideas, and critiquing claims.

Competency Builder 3.5
Conjectures and Counterexamples

When critiquing someone's argument, counterexamples can be used to prove their claim is not universally true, making them central to MP3. Additionally, you can build the skills of self-awareness and social awareness by engaging students in examining conjectures and counterexamples. Read aloud the book *Nothing Like a Puffin* by Sue Soltis. While this is a picture book, it can be used effectively with students at any grade level to explore the concept of counterexample. It begins by asserting that "there is nothing like a puffin," and compares the puffin to various objects. Each comparison produces examples and a counterexample. After reading the book aloud, reread and have students record each conjecture along with the counterexample on a table like the one in Table 3.2.

Table 3.2 • *Nothing Like a Puffin Conjectures and Counterexamples*

Conjecture	Counterexample
A newspaper is nothing like a puffin.	Newspapers and puffins are both black and white.
Jeans are nothing like a puffin.	
A goldfish is nothing like a puffin.	
A shovel is nothing like a puffin.	
A snake is nothing like a puffin.	
A helicopter is nothing like a puffin.	

online resources Download this table at https://companion.corwin.com/courses/wellroundedmathstudent

Extend this to incorporate self-awareness and social awareness by having students find connections between themselves and classmates. Ask students how understanding similarities and differences among us helps us in our everyday life and in the classroom. Provide them with a recording sheet with a table (Table 3.3) to compile comparisons between themselves and several classmates. Discuss things they might consider (other than gender), such as birthdays, birthplaces, hair color, hobbies, places they have visited, music preferences, extracurricular activities, etc. To complete the table, students fill in the name of the person they are talking to in the first column, provide an example (way they are different) in the second column, and give a counterexample (way they are alike) in the third column.

Table 3.3 • *Peer Examples and Counterexamples*

Conjecture	Example	Counterexample
_____ and I are nothing alike.		
_____ and I are nothing alike.		
_____ and I are nothing alike.		
_____ and I are nothing alike.		
_____ and I are nothing alike.		

online resources: Download this table at https://companion.corwin.com/courses/wellrounded mathstudent

Follow up with mathematics-specific conjectures aligned to the content standard. For example,

- Doubling a number always makes it larger.
- When you add numbers, they always get bigger.
- Fractions are always between 0 and 1.
- All shapes with 4 sides are rectangles.
- No square has a value for perimeter that is the same for the area.
- No prime numbers are even numbers.
- If a number is a perfect square, then it is not odd.

In summary, you have many choices in how to design a lesson with a focus on this mathematics goal and practice standard. It is important to recognize that argumentation may not be embedded in the task as it is presented, but rather in how the task is implemented in the classroom (Ayalon & Hershkowitz, 2018). This is also true for the integration of social-emotional competencies. We do not see those skills when looking at the paper version of the task, but rather we see it in the implementation of the task.

Assessing Interpersonal and Intrapersonal Skills

As you complete your planning, ask, **"How will I assess students' progress toward the mathematics goal of this lesson, their engagement in the**

mathematical practice standard, and their ability to use and continue to apply social-emotional competencies? How will I provide feedback? How will I build a chance for students to reflect on their intrapersonal and interpersonal skills?"

Recall the actions associated with constructing arguments and critiquing reasoning at the beginning of the chapter. MP3 and the social-emotional competencies used with it can be assessed in a variety of ways. First, observe students as they work with their peers for evidence of social awareness, decision-making, assertiveness, empathy, and/or self-awareness. Second, ask them questions to prompt explanations for how they use these skills. Third, engage students in self-assessment by providing them with prompts and asking them to reflect on how they used these skills. Consider the following prompts:

- Revisit "**Social Awareness** Skills" (Figure 3.1). What skills did you use?

- What **decisions** did you have to make? What helped you to make the decisions?

- In what way were you **assertive**? Provide at least one example. How did being assertive make you feel?

- How did you show **empathy** in today's lesson? Provide at least one example. How did others respond?

- How did you handle feedback, both positive and negative? (self-awareness)

- What did you or can you learn from the feedback you got from others? (self-awareness)

You can use recording sheets to collect assessment data as you listen to and observe students. Use a whole class observation tool (Table 3.4), an individual student observation tool (Table 3.5), or a self-assessment tool (Table 3.6) to capture evidence of students' development of their mathematical content knowledge, engagement in MP3, and the growth of their intrapersonal competencies as well.

Table 3.4 • *Whole Class Observation Tool*

Name	Mathematics Goal	Engagement in Practice Standard *Construct arguments*	Engagement in Practice Standard *Critique the reasoning of others*	Intrapersonal Competency *Self-awareness*	Interpersonal Competency *Social awareness*

Note: *Progress will be marked using 0–No evidence, 1–Little evidence, 2–Adequate evidence*

[online resources] Download this table at https://companion.corwin.com/courses/wellroundedmathstudent

Table 3.5 • *Individual Student Observation Tool*

Name of student:

Mathematics Goal	No Evidence	Little Evidence	Adequate Evidence
Engagement in the Practice	No Evidence	Little Evidence	Adequate Evidence
Constructs arguments			
Critiques the reasoning of others			

(Continued)

(Continued)

Social-Emotional Competencies	No Evidence	Little Evidence	Adequate Evidence
Self-awareness			
Social awareness			

Download this table at https://companion.corwin.com/courses/wellroundedmathstudent

Table 3.6 • *Self-Assessment Checklist*

Social-Emotional Competency	Not Sure	Not Yet	Getting There	Got It
Self-awareness				
Decision-making				
Social awareness				
Empathy				
Assertiveness				
Other skills used:				

Download this table at https://companion.corwin.com/courses/wellroundedmathstudent

LOOKING AT EXEMPLARS IN ACTION

Having explored the integration of content standards, mathematical practice standards, and social-emotional competencies, let's now examine specific examples from both elementary and secondary levels, focusing on the standards and competencies highlighted in this chapter.

Mr. Hernandez and His Fourth-Grade Class

Mr. Hernandez's fourth-grade class is working on "decomposing a fraction into a sum of fractions with the same denominator in more than one

CHAPTER 3 • GROWING SELF-AWARENESS AND SOCIAL AWARENESS

way, recording each decomposition by an equation" (4.NF.B.3b, CCSSM, 2010). They are expected to justify these decompositions using various representations. To access prior knowledge, Mr. Hernandez began with a class discussion about what it means to decompose a fraction into unit fractions. He assigns students to groups of four and provides each group with a set of fraction cards. He explains to the students that they are to represent the fraction on the card by decomposing it into unit fractions and then justify it by illustrating it with a fraction model. He provides specific instructions for how students are expected to work collaboratively. Working in small groups gives students an opportunity to share their decompositions, evaluate the work of their peers, and give/receive feedback. As you read the classroom vignette, identify how the teacher engages the students in MP3, while helping them develop self-awareness, social awareness, empathy, and assertiveness. Think about ways you do this in your own classroom.

Mr. Hernandez says, "Think about what it means to decompose a fraction into unit fractions. What is a unit fraction? Talk with your group and write down some examples."

Lonnie tells Amaya, "A unit fraction has a 1 in the numerator, like $\frac{1}{2}$."

Amaya adds, "It's like taking one piece from equal parts, like $\frac{1}{4}, \frac{1}{3}$, or $+\frac{1}{5}$."

Mr. Hernandez continues, "Now, think about how to decompose $\frac{3}{4}$ into unit fractions. Write down your idea."

After a minute, he says, "Share your ideas with your partner. Are they the same?"

Students compare answers: $\frac{1}{4}+\frac{1}{4}+\frac{1}{4}$.

"Today, you'll come up with more ways to decompose a fraction," Mr. Hernandez explains. "Use unit fractions and other fractions. Work on your own first, then we'll share ideas."

The class works quietly, then Mr. Hernandez instructs, "Share one example with your group. Use the tiles to show your idea and ask if they agree."

(Continued)

(Continued)

Pointing to his tiles Jamal says, "I broke $\frac{7}{10}$ into $\frac{5}{10} + \frac{2}{10}$. Do you agree?"

Zuri responds, "Yeah, it still makes $\frac{7}{10}$."

Kennady adds, "I made $\frac{1}{10} + \frac{3}{10} + \frac{3}{10}$ by grouping the tiles differently."

"That works," Jamal agrees.

Zuri shares, "I did $\frac{2}{10} + \frac{2}{10} + \frac{3}{10}$. It's different from yours but still $\frac{7}{10}$."

Nia says, "I broke $\frac{5}{10}$ into $\frac{3}{10} + \frac{2}{10}$. It's a different way."

Pointing to his tiles Zuri challenges, "But that's the same as mine, just in a different order. See how the groups look the same."

Nia pauses, then says, "Oh, I see. Here's another way: $\frac{1}{10} + \frac{2}{10} + \frac{2}{10} + \frac{2}{10}$."

"That's different but still $\frac{7}{10}$," Kennady confirms.

Mr. Hernandez joins them. "Nice work! Now, reflect on what you learned. How did this help you think about decomposing fractions?"

Jamal says, "I see now that I can decompose into more fractions, not just two."

Mr. Hernandez probes, "How would you break down $\frac{5}{10}$?"

Jamal replies, "I could do $\frac{2}{10} + \frac{3}{10}$, or even $\frac{1}{10} + \frac{1}{10} + \frac{3}{10}$."

Mr. Hernandez asks, "Anyone else see new ways to decompose fractions?"

Nia adds, "At first, I thought Zuri's and my decompositions were different, but they were the same. So, I tried a different way and broke $\frac{3}{10}$ into $\frac{1}{10} + \frac{2}{10}$."

Mr. Hernandez responds, "Zuri was assertive and disagreed with you. How did that feel?"

> Nia says, "At first, I was confused, but when we used the tiles, I understood. I'm glad he said something. It helped me see it differently."
>
> Mr. Hernandez smiles, "That's a great example of self-awareness." ●

Mr. Hernandez promotes **social awareness** by structuring the activity so all students contribute thinking toward accomplishing the learning goal, share their unique ideas, and show respect for ideas that may be different from their own. He explicitly states that different ideas and contributions of everyone exposes them to different perspectives, which enables them to develop **empathy**. Comparing their ideas to others' ideas promotes both **social and self-awareness**. In addition, sharing their ideas promotes **assertiveness** and receiving feedback from peers which also promotes **self-awareness**.

Mx. Adams and Their High School Algebra Class

Mx. Adams's mathematics lesson is focused on solving "quadratic equations by inspection, taking square roots, the quadratic formula and factoring, as appropriate to the initial form of the equation" (HS.REI.B.4b CCSSM, 2010). They recognize that this standard involves teaching students how to use a particular method and when to choose one. This lends itself to MP3 because students can make an argument to justify the method they choose. Mx. Adams creates a master list on the board of methods students have learned that includes factoring, taking square roots, graphing, and using the quadratic formula, and then tells them the goal today is to choose a method based on the equation's features. Mx. Adams created 24 quadratic equations, writing each on a notecard, numbered them one through six, and repeated the numbering four times. They then handed a notecard to each student and asked them to take out a piece of notebook paper.

Mx. Adams explains that they will work on both self-awareness and social awareness during the activity today and reviews what those words mean. Explaining that the activity might feel difficult in the beginning, Mx. Adams notes that students will gain more confidence the more they practice choosing a method. Mx. Adams tells the students they want them to be aware of how their confidence changes as they practice choosing methods and getting feedback throughout the lesson. They ask students to be mindful of how they interact with each other, encouraging them to be assertive by sharing their ideas and offering feedback to their peers, while also emphasizing the importance of showing empathy toward one another.

Mx. Adams says, "Look at the quadratic equation on your notecard. Decide which method seems best for solving it and note why. You don't need to solve it yet, just pick your method. Take a moment to observe the equation's structure to help guide your decision."

After two minutes, they continue, "It's okay to feel unsure. We can often make better choices with more information, so stay open to other methods. You'll now work with others to explore different options. If you hear a new idea, feel free to change your method. Go to the number on your notecard and find your spot in the room."

Once the students are in their places, Mx. Adams explains, "Sharing our ideas helps us understand each other's thinking. Form a circle, hold up your notecard, and explain your method. Listen respectfully, and after each person shares, two others should give constructive feedback. Here are some prompts to guide your feedback:

- I agree because . . .
- I noticed . . .
- I thought about . . .
- Another way to do this is . . ."

In Group 1, Paulo says, "I chose graphing for $(x-4)^2 = 25$ because I couldn't factor it."

Lucy responds, "I agree; it doesn't have a middle term, so graphing makes sense."

Karmen adds, "You could also square root both sides since you have a binomial squared on one side."

Paulo agrees, "That's a good point—it's easier than graphing."

Lucy continues, "I also chose graphing for a similar reason; it didn't seem easy to factor."

CHAPTER 3 • GROWING SELF-AWARENESS AND SOCIAL AWARENESS

Paulo replies, "I think graphing works, but you could also use the quadratic formula."

Once the sharing is complete, Mx. Adams directs the students to return to their seats and solve the equation. They monitor the room as students work.

Mx. Adams continues, "You've solved the equation in one way. Now try solving it using a different method."

A few minutes later, Mx. Adams asks, "Talk to your partner about the two methods you used and which one you preferred. Remember, your partner's opinion isn't about being right or wrong—it's about their preference. Keep an open mind!"

Paulo shares, "At first, I chose graphing because factoring seemed hard, but when someone suggested square rooting, it seemed easier. I tried both methods, and I realized I missed the fact that it could be +5 or -5 when I used the square root."

Yolanda responds supportively, "I can see that happening! It's important to check if there's one solution, two, or none. I think both methods are useful. The square root method is quick if you remember the plus/minus."

Paulo reflects, "Now I know to check for both solutions. I prefer graphing—it's easier to see all the solutions."

Yolanda shares her methods, and the conversation continues. ●

This lesson provided the opportunity to explicitly address **social awareness**. Students can demonstrate **assertiveness** as they share and provide constructive feedback. They do not have to agree with the person's choice but need to have **empathy** and consider the person's feelings as they give feedback. Mx. Adams provides structures to enable students to provide feedback to peers that is effective and supportive. This lesson also lends itself to an opportunity for students to develop **self-awareness** as they reflect on their feelings and thoughts as they share and explore their ideas and justify their conclusions. Students will find themselves challenged by the critiques of their peers, which supports the development of a growth mindset.

Reflection

Revisit Table 3.1 from the introduction of this chapter. In what ways did these teachers address the two aspects of this practice: construct viable arguments and critique the reasoning of others? What role did social-emotional competencies play in supporting these actions? Across both classroom narratives, the teachers intentionally plan and prepare for the integration of content standards, practice standards, and intrapersonal and interpersonal learning competencies—specifically, self-awareness, social awareness, empathy, and assertiveness. The teachers in these narratives do this with thoughtful and purposeful pedagogical choices while utilizing a variety of instruction tools and methods. Specifically, they provide structures to support students in interacting in ways that allow all to contribute. They designed lessons to provide opportunities for students to agree or disagree with each other or prompt their classmates to consider other choices. They make explicit the social-emotional aspects that are integrated into the lesson.

SUMMARY

In this chapter we reviewed the three key elements needed in building lessons that incorporate mathematics with social-emotional competencies and offered two classroom narratives that utilize MP3 to illustrate this process. We shared ways to explicitly address this practice while simultaneously incorporating social-emotional competencies. We illustrated how argumentation supports the development of students' mathematical autonomy and strengthens their confidence by giving them ownership of the mathematics they are learning. We demonstrated that students could generate their own ideas and, if they are open-minded, can improve on these ideas based on feedback from peers. Additionally, we provided routines to encourage students to listen to others' ideas, consider if they make sense, and offer feedback in a respectful, productive manner.

Now we will consider specific questions to think about and actions to take that promote the development of social-emotional competencies that can enhance the teaching and learning of mathematics.

Questions to Think About

1. In what ways did I model how to build a strong argument, making a clear connection between the evidence and a claim? How did I model and reinforce the importance of respecting others' ideas, even if students disagreed with them?

2. What strategies do I use to encourage students to develop and express their own mathematical arguments?

3. How did I give students opportunities to practice empathy by considering how others might feel when their ideas are critiqued or questioned? In what ways do I foster an atmosphere where students can support each other's reasoning and provide constructive feedback?

4. Reflect on a recent mathematics lesson you taught, then ask yourself, How did I create a classroom environment where students felt comfortable sharing their arguments without fear of judgment?

5. How do I encourage students to stay persistent and focused when constructing arguments or critiquing others' reasoning?

Actions to Take

1. Share common mistakes by providing an incorrectly worked example and asking students to examine their work to see if this was a mistake they made. Present the example, tell the students there is a mistake, and ask them to find it. Have them write a response for how they would explain the mistake to a classmate.

2. Identify when students need to engage in conflict management in a pair or group. Ask students to describe from the other person's perspective the challenge they are experiencing.

3. After a mathematics lesson (or series of lessons), have students help you create a list of the skills they have been working on, identify something they feel confident doing, and complete the statement, "I can . . ."

4. Implement language frames and model the use of the language.

CHAPTER 4

PROMOTING DECISION-MAKING WHEN MODELING WITH MATHEMATICS

DOROTHEA BRANDE, author of a bestselling self-help classic *Wake Up and Live!*, once said, "A problem clearly stated is a problem half solved." Her book offers practical advice on developing a proactive, positive mindset, and her quote emphasizes the importance of clearly defining and understanding a problem before attempting to solve it. When a problem is clearly articulated, it becomes easier to identify the underlying issues, explore potential solutions, and develop a strategy to address it. This clarity helps eliminate confusion and directs efforts more efficiently, making the resolution process much smoother and more effective. Although Brande emphasized personal development and urged readers to conquer self-doubt and embrace life, her quote can just as effectively be applied to the practice of mathematical modeling. Mathematical modeling is widely used in various fields, such as engineering, economics, biology, and environmental science, where it helps in understanding complex systems and making informed decisions, but it can also be used to better understand and solve problems in everyday life. Problem identification and problem formulation are the first two steps of the modeling process.

> **MODEL WITH MATHEMATICS**
>
> *Mathematically proficient students can apply the mathematics they know to solve problems arising in everyday life, society, and the workplace. In early grades, this might be as simple as writing an addition equation to describe a situation. In middle grades, a student might apply proportional reasoning to plan a school event or analyze a problem in the community. By high school, a student might use geometry to solve a design problem or use a function to describe how one quantity of interest depends on*
>
> *(Continued)*

> (Continued)
>
> *another. Mathematically proficient students who can apply what they know are comfortable making assumptions and approximations to simplify a complicated situation, realizing that these may need revision later. They are able to identify important quantities in a practical situation and map their relationships using such tools as diagrams, 2-by-2 tables, graphs, flowcharts, and formulas. They can analyze those relationships mathematically to draw conclusions. They routinely interpret their mathematical results in the context of the situation and reflect on whether the results make sense, possibly improving the model if it has not served its purpose. (CCSSM, 2010)*

Mathematical modeling is a process that allows students to gain deeper insight into their world and recognize mathematics as one tool they can use to solve problems and make informed decisions. It involves gathering information to create mathematical representations of real-world situations that provide a foundation for understanding, analyzing, predicting, explaining, and solving problems "arising in everyday life, society, and the workplace" (CCSSM, 2010). The focus on identifying solutions to personal and social problems also makes it a powerful practice for developing social-emotional competencies. Mathematical modeling relies heavily on responsible **decision-making**. Modeling requires the modeler to ask questions, make judgments, and be open-minded about what to include or exclude when creating a model. According to Arnold et al. (2021), judgments reflect values and "because making decisions about what is important involves seeing situations from others' perspectives, modelers use **empathy**" (p. 7).

While mathematical modeling and modeling mathematics sound synonymous, there is an important distinction between them that can shed more light on mathematical modeling and its development. Mathematical Practice 4 (MP4) focuses on mathematical modeling, but many content standards across grade levels incorporate modeling mathematics. In modeling mathematics, models are conceptual structures used to represent and relate aspects of mathematical situations. These conceptual structures, or representations, can be classified into five general types: physical (manipulative models), visual (pictures), verbal (oral language), contextual (real-world situations), and symbolic representation (written symbols) (Lesh et al., 1987; Van de Walle, 2007). According to Cirillo et al. (2016), "modeling mathematics refers to using representations of mathematics to communicate mathematical concepts or ideas. A key feature of modeling mathematics is that the process *begins in the mathematical world*, rather than the real world" (p. 4). Using models to represent mathematics (modeling mathematics) is important because it provides access to the mathematical concept, and it lays a foundation for developing mathematical modeling skills. In other words, models are used as tools to help students make sense of concepts and solve mathematical problems. This is addressed further in the discussion of MP2: Reasoning abstractly and quantitively and MP5: Use appropriate tools strategically.

On the other hand, mathematical modeling (MP4) *begins outside of mathematics* with a real-world, authentic situation and a process of mathematizing the problem by creating a model and then reconciling the model with the real situation (Cirillo et al., 2016). Mathematical modeling is an active and creative process of constructing, testing, revising, and ultimately applying a conceptual representation within a context. A model in this case is the outcome of the modeling process. As illustrated in Figure 4.1, modeling is a cyclical process that starts with a real-world situation and requires students to identify and define a problem, formulate a real-world model, mathematize to create a mathematical model, perform mathematical work to achieve a result, interpret and recontextualize the result to create real-world meaning, and then validate the result with the real-world situation (Kaiser & Stender, 2013).

Figure 4.1 • *A Representation of the Modeling Cycle*

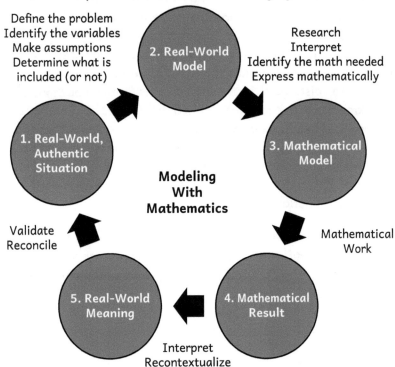

You can leverage MP4 to develop intrapersonal and interpersonal skills necessary for problem-solving in math class and beyond. Engaging with others to model with mathematics promotes the development of **responsible decision-making**, **social awareness**, **creative thinking**, and **adaptability**.

MERGING CONTENT STANDARDS, MATHEMATICAL PRACTICES, AND SOCIAL-EMOTIONAL COMPETENCIES

As we've seen in the other chapters, there are numerous decision points when planning and preparing mathematics lessons that intentionally integrate MP4 and social-emotional competencies. Let's go back to our framework to see how we plan to deliberately integrate the social-emotional competency skills of **responsible decision-making**, **social awareness**, **creative thinking**, and **adaptability** within a mathematics lesson that leverages MP4.

Mathematical Content Standard and Corresponding Mathematics Goal

Planning for a lesson begins by asking **"What is the math goal of this lesson?"** Consider a fourth-grade standard: "apply the area and perimeter formulas for rectangles in real world and mathematics problems" (4.MD.3, CCSSM, 2010). A goal of a lesson involving this standard would focus on students understanding the meaning of area and perimeter and then identifying situations in the real world that involve finding area and perimeter to apply the appropriate formula. At the secondary level, students learn to "distinguish between situations that can be modeled with linear functions and with exponential functions" (F.LE.1, CCSSM, 2010). In the real world, students are regularly exposed to contexts that can be modeled with linear, quadratic, and exponential functions (and other functions). One lesson goal related to this standard would focus on students identifying what family of functions best models a particular situation and then generating a function.

It is important to note that establishing mathematical goals for a modeling lesson often happens differently from planning other math lessons. One major step in the modeling process is deciding what mathematics the situation requires and, as a result, one can only anticipate what content standards may align. Two broad questions can be used to guide goal setting for modeling lessons (Arnold et al., 2021):

1. As an outcome of engaging in mathematical modeling, what should students be able to do?
2. What perspectives about themselves, mathematics, and their communities should students develop?

Reflecting on the fourth-grade standard, our goal is for students to see how area and perimeter are relevant in everyday life, helping them connect mathematics to their personal experiences. For the high school standard, we aim to go beyond simply generating functions; we want students to use them to gain insights into varying situations and better understand their changing world.

Mathematical Practice

Next, reexamine the language in the standards and goals to recognize how it aligns to the math practices. Ask, **"Which mathematical practice supports engagement in this content standard?"** There are many standards across grades K–7 that use the language "solve word problems" or "real-world problems." Additional examples include the following:

- Use addition and subtraction within 20 to solve word problems involving situations of adding to, taking from, putting together, taking apart, and comparing with unknowns in all positions, e.g., by using objects, drawings, and equations with a symbol for the unknown number to represent the problem. (CCSSM, 1.OA.A.1)
- Use ratio and rate reasoning to solve real-world and mathematical problems, e.g., by reasoning about tables of equivalent ratios, tape diagrams, double number line diagrams, or equations. (CCSSM, 6.NS.1)

Standards like these allow students to engage at some level in MP4. Students decontextualize word problems so that mathematics can be used to solve these problems, and the solution must be recontextualized and interpreted. In this way, MP4 connects with MP2: Reason abstractly and quantitatively. However, word problems are not considered equivalent to mathematical modeling because word problems are often clearly presented, involve little to no extraneous information or decision-making, and do not require students to evaluate, revise, or improve their solutions (Lesh & Harel, 2003). Word problems are over-simplified and often unrealistic, while modeling situations are more authentic and relatable. Word problems do, however, require students to engage with aspects of the modeling cycle, providing students an opportunity to develop foundational skills. Solving a word problem or a real-world application problem does not mirror the same complexity of mathematical modeling, but it does introduce students to the idea that problems exist in the real world (outside of mathematics) that can be interpreted and expressed mathematically. With this in mind, we can take steps to transform a word problem to a modeling problem by opening the problem (see Figure 4.2).

Figure 4.2 • *Transforming a Mathematical Problem or a Word Problem Into a Modeling Problem*

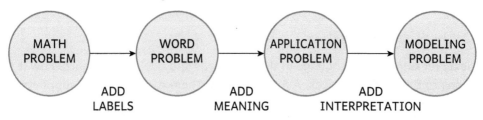

Source: Garfunkel and Montgomery (2019). View the entire report, available freely online, at https://www.siam.org/Publications/Reports/Detail/guidelines-for-assessment-and-instruction-in-mathematical-modeling-education

The process involves more than putting the math problem into context, but rather giving it meaning and opening it up to interpretation and to making choices. For example, consider the mathematical problem in Table 4.1.

Table 4.1 • *Transforming a Mathematical Exercise Into a Modeling Problem*

Math Exercise	Word Problem	Modeling Problem
4 + 3 = ?	Leo has 4 cookies. Deb has 3 cookies. How many do they have all together?	After lunch, you and a friend would like a few cookies for dessert. Talk with your partner and determine how many cookies you and your friend should ask for.

Source: Adapted from Garfunkel and Montgomery (2019). View the entire report, available freely online, at https://www.siam.org/Publications/Reports/Detail/guidelines-for-assessment-and-instruction-in-mathematical-modeling-education

First, "math problem" was traded for "math exercise," because it is considered a skill. A true problem does not have a clear path for solving it, and it involves investigation and problem-solving. Notice how in Figure 4.2, transforming from a math "problem" to a word problem involves little more than adding labels to the numbers in the equation, so it still feels very much like an exercise. However, the modeling problem creates an authentic opportunity for students to interpret the problem (i.e., what does "few" mean) and make choices. Talk with students about what the word *should* implies in this situation. How is what you "should" ask for different from what you "want to" ask for in this situation? In this way, responsible decision-making, social awareness, and self-awareness are evident in modeling problems. Unpacking the context enables students to move from a real-word authentic situation to a real-world model as they define the problem and determine what they need to know. They then express this mathematically using a mathematical model, do the math, and put it back into context to respond to the situation. This shows how students as young as kindergarten and first grade can engage in mathematical tasks that allow them to develop interpersonal and intrapersonal skills.

At the high school level, the connection between content standards and mathematical modeling is made explicit. Modeling is considered its own conceptual category but does not have separate standards. Rather, some standards are marked with a star symbol within the Algebra, Function, and Geometry conceptual categories to indicate that these are modeling standards and can be applied meaningfully to real-world modeling situations. The entire Statistics and Probability conceptual category is marked with a star indicating that all the standards in the category are considered modeling standards. Students are expected to learn to use statistics and probability to analyze and understand empirical situations, apply them to work and everyday life, and utilize them to improve decision-making.

Social-Emotional Competencies

Based on what it means to be engaged in MP4, ask, **"What intrapersonal and interpersonal skills are inherent, are needed, and can be further developed for students while engaging in MP4?"** Whether they are word problems, real-world problems, or mathematical modeling problems, it is essential to acknowledge that "[p]roblem-solving is not only a quantitative challenge but a human one. We can engage students in deeply understanding what resolving problems entails and what it feels like, which is satisfying for children and adults alike" (University of Texas at Austin, n.d.). When presented with a modeling context, students find themselves face-to-face with the complexities of reality. They are called on to make many decisions and are faced with the challenge of determining what information they have and do not have. What information do they need and not need? This presents them with opportunities to ask questions, make assumptions, seek information, apply prior knowledge, learn new skills, use critical and creative thinking to arrive at a solution, and then determine whether it is reasonable in the given context. This requires students to draw on, but also continue to develop, many intrapersonal and interpersonal skills, but relies heavily on **responsible decision-making, social awareness, creative thinking, and adaptability.**

Intrapersonal Skills

Responsible decision-making is intrapersonal in nature when seen as making a choice by using one's judgment. As early as two years old, children can learn to make choices. Giving them choices provides an opportunity to make decisions, recognize the consequences of the decision, and feel more in control. However, making choices is quite different from the decision-making required in a modeling situation. In a modeling situation, students are generating the options rather than being given options. Research indicates that young children have difficulty collecting information, ignoring irrelevant information, and attending to relevant information (Davidson, 1991; Howse et al., 2003). Training, experience, and maturity enable students to overcome some of these difficulties.

Modeling mathematics involves **creative thinking** as students develop original ideas for solving problems. Creating models to represent mathematical problems is an essential part of the problem-solving and modeling process (Hartono, 2020; O'Connell & SanGiovanni, 2013). Students brainstorm ideas and initially explore them using concrete referents, such as objects, drawings, diagrams, and actions. Over time, these ideas become more abstract in nature. Modeling situations require students to look at ideas from multiple perspectives, take risks, and try new approaches.

Interpersonal Skills

The modeling process involves returning to previous stages as new information is received and new questions arise to incorporate what is learned into the model. Students need to approach modeling flexibly and make adaptations as they proceed. Modeling often involves interaction with others. When students participate in and contribute to modeling activities in a group setting, it gives them a sense of belonging, not only in the math classroom, but in the community and the world. In this way, real-world, authentic situations allow students to build **social awareness** as they compare different approaches based on the different perspectives of their classmates. This leads to social decision-making, which combines both the decision-making process and social awareness. **Social decision-making** is influenced by the interaction between emotion and cognition (Luo & Yu, 2015). As children mature, so do the cognitive and emotional abilities that enable them to consider the perspectives of others and to analyze and interpret more complex issues. However, there are contexts in which emotion overrides reason in the decision-making process, such as when cognitive capacity is weak (e.g., incomplete information, limited decision time, and impaired self-regulation) or emotion is strong (e.g., threatening cues, self-related information, and social stimuli) (Luo & Yu, 2015). This is particularly important to consider in preadolescence when children are likely to have increased feelings of independence, desire more autonomy, and begin to take on greater responsibility. Preadolescent and adolescent children feel an increase in the pressure to "fit in" as the brain is growing more sensitive to social cues resulting in greater emotion. It is important that teachers provide opportunities for students to reflect on their feelings, thoughts, ideas, and actions during and after modeling experiences.

Instructional Structures and Engagement Strategies

Now shift toward planning and ask, **"With an eye on our math goal, how will I support social-emotional development as I engage learners in MP4: Model with mathematics? What structures, strategies, methods, and/or tools can I use?"**

The complexity of the modeling process is time demanding, so modeling activities are spread out over the school year. Students typically engage in modeling activities in cooperative groups in a classroom setting. Depending on the problem students are presented with, a modeling activity would take an entire class period (60–90 minutes) or several class periods to complete. During the lesson, structures are in place as questions and prompts to guide students through the modeling process. After students have solved the problem and created an appropriate and useful model, they share their solution with the class. It is important that students prepare to communicate their model with the class. The presentation, and discussion associated with it, should lead students to revisit their model to revise and improve it.

Focusing on Decision-Making Approaches

Modeling math relies heavily on decision-making. We have long known "the way a child learns how to make decisions is by making decisions" (Kohn, 1993). Miller (2023) states there are basic ways to promote decision-making, which starts with a decision that affects children and then letting them take part in the decision-making process. Ask questions like "What are the options? What are the benefits and advantages or limitations and disadvantages of these options? How will this decision affect others?" According to the Life Skills Group (2018) there are five ways to improve decision-making, and we can think about how these can be applied to math class.

> **Competency Builder 4.1**
> *Identifying Problem-Solving Approaches*
>
> Begin by explicitly teaching the four steps of decision-making, which include the following:
>
> 1. Identify the decision being made. (What am I trying to find? What information do I have?)
> 2. Identify the options. (What methods can I use? What math do they need?)
> 3. Evaluate the options. (What's the best one?)
> 4. Based on this evaluation, decide on the best method and solve the problem.
>
> A graphic organizer like the one in Figure 4.3 is a helpful way to organize these steps and translate among them.
>
> **Figure 4.3** • *Organizing Your Decision-Making*
>
> | Problem or Question: | |
> |---|---|
> | Identify the decision(s) you need to make. | Brainstorm methods for solving the problem and needed information. |
> | Evaluate the options. List the strengths and weaknesses of each option. | Explain the reason you selected this method. |
>
> **online resources** Download this figure at https://companion.corwin.com/courses/wellroundedmathstudent

Next, involve students in decision-making, which is especially important with modeling because there are many factors to include or exclude. Provide opportunities for students to practice making decisions. Finally, teach them to ask questions of themselves, such as "What makes this a good choice? Why does this work? Is this the best option?" It is important that students reflect on the outcome of their decision and seek feedback and recommendations. The most important action you can take is to value the time and thought that goes into a decision.

Using Physical Models to Explore Geometric Relationships

Use manipulatives, such as geoboards, attribute blocks, tangrams, multi-link cubes, and geometric shape sets, to enable students to explore geometric concepts including attributes of shapes, geometric patterns, symmetry, similarity, 2D and 3D measurement, and geometric transformation. Have students experiment with creating different shapes and combining shapes to create new ones. For example, they can create a model of an ice cream cone using a cone and a hemisphere with the same base circle. Manipulating concrete models will allow students to meaningfully adapt formulas and apply principles to model and solve problems in situations that are new to them. This fosters skills of creative thinking and adaptability.

Competency Builder 4.2
Building and Analyzing Shapes

Provide students with a set of tangrams. Introduce them by explaining what they are and showing students the seven pieces. Tell them to take one minute to notice two or three relationships among the pieces. Have them share what they notice. There are many things they may suggest. For example, the medium triangle, square, and parallelogram can be formed from two small triangles if they rotate the small triangles in different ways. There are many similar angles within these shapes and the angles are multiples of 45 degrees.

Next, explain that they are going to use all seven pieces to form a specific figure (e.g., a square, house, boat, dog, etc.) as seen in Figure 4.4. Many puzzles are available for download online. As they work, encourage them to try rotating, flipping, and rearranging the shapes. This exposes students to geometric transformation and encourages flexibility. As they create different shapes with

(Continued)

(Continued)

the same set of seven shapes, students are challenged to adapt their thinking to create a new shape using a different spatial arrangement. Not only do students build better spatial adaptability as they shift perspective, but they also adapt their approach.

Figure 4.4 • *Sailboat Tangram Puzzle*

Source: istock.com/apodiam

Specifically, ask students what strategies they use to adapt their thinking. They may suggest they use trial and error, or that they experiment by arranging shapes one way and then manipulate the shapes in different ways to see if they can get a better fit. They may suggest they look for patterns or relationships within the puzzle they are solving and imagine how the puzzle could be decomposed in different ways. End by asking them how they can use this flexible and adaptive thinking in everyday life.

Creating Opportunities to Ask Questions and Formulate Problems

At all levels, students must experience authentic problems. Authentic problems that lend themselves to modeling situations arise when there is no immediate and clear path for solving the problem. Modeling situations can be based on real-life scenarios found in our everyday world as well as urgent problems occurring on a larger scale. The modeling process starts with a real-world situation that requires students to identify and define a problem.

Therefore, planning for a modeling lesson begins by identifying, selecting, or creating high-quality mathematical tasks aligned to the standards. Problems must be open-ended, require students to find various solutions, share and evaluate the validity of the solutions, and then go on to revise and improve their models. We can generate our own problems by listening to students talk and collecting questions that they ask.

Competency Builder 4.3
Notice and Wonder Routines

Provide opportunities for students to initiate modeling tasks. Rather than give the student a real-world word problem, present the context and have students formulate the question. Learning to ask questions has been called "the ultimate learning skill" (Brodsky, 2021). Generating questions also promotes creative thinking and adaptability. Use Notice and Wonder routines to prompt a modeling task by presenting students with an image of a situation or setting and inviting them to share what they notice and wonder. Tell students to put on their creative thinking caps, share a picture or image, and ask them what they notice. For example, consider Figure 4.5, which depicts two avocados with vastly different-sized pits.

Figure 4.5 • *Avocado Notice and Wonder*

Looking at the images of the avocados, students notice the difference in the size of the pits and how one avocado has more "meat" than the other. They wonder if either of these are "normal" sized avocado pits. How much meat is there in a typical avocado? Either the teacher or students then select a notice or wonder to examine further through a mathematical lens and pose mathematical questions.

(Continued)

(Continued)

To make engagement with social-emotional competencies explicit, ask students to reflect on how they used creative thinking and adaptability in this routine. In this way, Notice and Wonder routines empower students by drawing on their insights, cultivating curiosity, and engaging them in critical thinking (Rumack & Huinker, 2019). Learning to ask mathematical questions is not only an essential aspect of the modeling process, but also an essential part of what it means to "do math." Students are also more motivated to search for answers to their own questions.

Competency Builder 4.4

Exploring Three-Act Math Tasks

The Three-Act Math Task uses multimedia storytelling to engage students in authentic and engaging problem-solving and mathematical modeling. It begins by evoking curiosity in students, leveraging that to lead them through problem posing, and engaging them in problem-solving. Tell students that they will need to use their decision-making and creative thinking skills in this lesson. Prompt them to pay attention to the ways they do this because you will ask them about it when they are done. Students develop creative thinking as they pose problems, seek out information, make choices, and apply their ideas to solve problems.

In Act 1, students are shown a short video clip that introduces the problem with the purpose of prompting students to notice, wonder, and ask questions (Meyer, 2013a). For example, consider this image of the end of a video revealing the sugar in a bottle of root beer (Figure 4.6).

Figure 4.6 • *Sugar Cubes Act 1*

Source: Fletcher (2023).

After showing the video, ask students what they notice and wonder. They may wonder if all soda has this much sugar or if diet soda would have less. As students close in on the details of the task, they might wonder how many grams of sugar that would be. They may wonder how many cubes are in the 2-liter bottle of root beer. Ask students for an estimate of how many sugar cubes are in the 2-liter bottle of root beer. Discuss estimates that are too high or too low. Ask what information they need to answer the question.

In Act 2, additional multimedia is used to provide additional information students can use to solve the problem (Meyer, 2013b). For example, in the Sugar Cubes task, share information about the grams of sugar in a cube and the grams of sugar in a 2-liter bottle (Figure 4.7).

Figure 4.7 • *Nutrition Labels for Sugar Cubes and Root Beer*

Source: Fletcher (2023).

In Act 3, a final video is provided that concludes the story and provides a potential solution to the problem and an opportunity for comparison and reflection (Meyer, 2013c). In the Sugar Cubes task, the final reveal shows 108 sugar cubes are in the 2-liter bottle (Figure 4.8).

(Continued)

(Continued)

Figure 4.8 • *The Final Reveal of the Sugar Cubes Task*

Source: Fletcher (2023).

Using the graphic organizer in Figure 4.3, encourage students to think about what they need to know and different ways they can solve the problem. Encourage the use of drawings, diagrams, or objects to model the problem. In this way, students practice translating the situation into a mathematical model, solve the problem, and put the solution back into context.

Summarize by asking students to work with a partner to list the ways that viewing and solving problems in this way helps them to expand their social awareness. Ask them to reflect on how they used new information they collected along the way. Discuss how asking questions and gathering information enables them to make responsible decisions and to adapt their thinking as new information becomes available. Three-Act tasks enable students to "pose their own mathematical questions and reason about the world around them, wade through the messiness of real life for needed information, to model situations using the mathematics they have learned, to predict and estimate, and to be driven to engagement by curiosity" (Englard, 2015). In general, encouraging students to notice the world around them promotes social awareness, and learning to ask important questions is an crucial life skill.

Some Sources for Video Versions of Word Problems

3-Act Task File Cabinet (Fletcher, n.d.). (https://gfletchy.com/3-act-lessons/)

3 Act Tasks (Acosta, n.d.). (https://kristenacosta.com/3-acts/)

Video Story Problems (Rimes, 2012)
(https://www.techsavvyed.net/archives/2352)

Robert Kaplinsky Real World Lessons (Kaplinsky, 2016)
(https://robertkaplinsky.com/lessons/)

3 Act Math Tasks by numerous authors
(https://tapintoteenminds.com/3act-math/)

Using Children's Literature to Create Modeling Activities

Modeling activities often address more than one mathematical concept and require students to engage in problem-solving, reasoning, and communication. Activities that stem from literature, like the one shared in Competency Builder 4.5, simultaneously leverage and develop responsible decision-making, social awareness, creative thinking, and adaptability.

Competency Builder 4.5
Math Curse

Math Curse (Scieszka & Smith, 1995) is a book about a student who feels trapped in a "math curse" after hearing the teacher say, "You know, you can think of almost anything as a math problem." The book presents everyday situations that students can relate to and allows them to connect mathematical concepts to real-world problems, a key aspect of mathematical modeling. The narrative encourages students to approach problems creatively. As the girl in the book encounters various math-related challenges, students can be prompted to explore different representations for the same problem, enhancing their modeling skills. The exaggerated scenarios invite students to identify the variables and assumptions involved, analyze and critique the reasoning used, and refine the models. Each scenario serves as a springboard for students to work together to create and refine mathematical models based on what is presented in the book.

Read the book aloud then invite groups of students to select a scenario to investigate. Revisit the modeling cycle (Figure 4.1) with students and discuss how they can employ the cycle as they explore their chosen scenario. They will do the following:

1. Define the problem embedded within the scenario. They will identify and record the variable, make assumptions, and determine what information they need to include (or not include).

(Continued)

(Continued)

> 2. Collect information that is needed and express the situation mathematically.
> 3. Do the math.
> 4. Interpret the results of the work back in the context of the situation selected to see if it makes sense.
> 5. Explain how they would revise the original model.
>
> Have students record their ideas at each step on a large sheet of paper. Facilitate a class discussion where groups share what they did in each step of the modeling cycle. Culminate by discussing how they would amend the scenario and create a revised edition of the story. To focus on the development of social-emotional competencies, ask students to share ways that they employed creative thinking as they explored the scenario. Revisit Competency Builder 3.1, specifically Figure 3.1 "Social Awareness Skills," and ask students to identify skills they demonstrated in this activity.

Assessing Interpersonal and Intrapersonal Skills

Last, ask, **"How will I assess students' progress toward the mathematics goal of this lesson, their engagement in the mathematical practice standard, and their ability to use and continue to develop social-emotional competencies? How will I provide feedback? How will I build a chance for students to reflect on the development of their intrapersonal and interpersonal skills?"**

Recall the discussion of mathematical models and modeling with mathematics at the beginning of the chapter. All the competency builders provide opportunities for you to observe students, listen to them, and encourage them to reflect on the impact of their thoughts, feelings, and actions on their learning and that of others. To capture evidence, you can monitor individual students, observe the work of small groups, and facilitate whole class discussions. As you do, listen and watch for evidence of student actions that demonstrate they are engaging with the modeling process and growing in their social awareness, decision-making, creativity, and adaptability. You can record evidence in a whole class observation tool (Table 4.2) or an individual student observation tool (Table 4.3) to document students' development of not only their math content knowledge and engagement in MP4 but the growth of their intrapersonal competencies as well.

Table 4.2 • *Whole Class Observation Tool*

Name	Mathematics Goal	Engagement in Practice Standard *Decontextualizes to generate math model*	Engagement in Practice Standard *Does the math work and puts back into context*	Intrapersonal Competency *Responsible decision-making*	Interpersonal Competency *Social awareness*

Note: *Progress will be marked using 0–No evidence, 1–Little evidence, 2–Adequate evidence*

Table 4.3 • *Individual Student Observation Tool*

Name of student:

Mathematics Goal	No Evidence	Little Evidence	Adequate Evidence
Engagement in the Math Practice	**No Evidence**	**Little Evidence**	**Adequate Evidence**
Decontextualizes to generate math model			
Does the math work & puts back into context			

(Continued)

(Continued)

Social-Emotional Competencies	No Evidence	Little Evidence	Adequate Evidence
Responsible decision-making			
Social awareness			

Question prompts can be used as you interact with students and can be presented to students as self-assessment prompts. A self-assessment checklist is provided in Table 4.4. Many prompts are embedded within in the competency builder activity description, but consider the following prompts:

- Look at the list of skills associated with social awareness (i.e., Figure 3.1 in Chapter 3). Share two examples of how you used these skills to improve your **social awareness**. What was the outcome of using those skills? If the results were not positive, how could you reframe your thinking to make them positive?

- What **decisions** did you have to make? What helped you to make the decisions?

- How did you **adapt** your thinking or your approach to solving the problem? Provide at least one example.

- How did you practice **creativity** in today's math activities? Provide at least one example.

Table 4.4 • *Self-Assessment Checklist*

Social-Emotional Competency	Not Sure	Not Yet	Getting There	Got It
Social awareness				
Decision-making				
Creativity				

Adaptability				
Other skills used:				

LOOKING AT EXEMPLARS IN ACTION

Now that the merging of content standards, mathematical practice standards, and social-emotional competencies has been explored, let's look more closely at an elementary and a secondary example, using the standards and competencies focused on in this chapter.

Mr. Baker and His Fourth-Grade Class

Mr. Baker planned a lesson that addresses the standard "apply the area and perimeter formulas for rectangles in real world and mathematics problems" (CCSSM, 4.MD.3). He recognizes that if he is thoughtful about it, he can connect this standard to modeling with mathematics. He identifies several social-emotional competencies he wants to address. Specifically, **decision-making** and **social awareness**. The dialogues shared in this narrative reflect the beginning of the lesson where Mr. Baker sets the tone for the class. As you read the narrative from this fourth-grade classroom, identify how the teacher engages the students in MP4: Model with mathematics, while helping them develop responsible decision-making, creative thinking, social development, and adaptability. Think about ways you do this in your own classroom.

Mr. Baker begins by asking students to briefly share in their small groups any experiences they have had with sandboxes. After a minute or two, he displays some pictures of sandboxes of various kinds. He then poses the problem:

> **Pedro's dad wants to make a sandbox in the backyard for his little brother. What does he need to think about to build a sandbox?**

Mr. Baker tells the students they will work in their group to create a list of things that Pedro's dad needs to consider. To be sure that everyone has an opportunity to contribute and to invite multiple perspectives, Mr. Baker tells

(Continued)

(Continued)

them to go "round robin" around the group and have each person share one idea and continue going until there are no new ideas. Then, he goes directly to a group with a student that tends to dominate conversations. He has a plan to strategically have that student share last, so others have a chance to talk. To help this student build active listening skills, he asks the student going last to record the ideas for the group.

Mr. Baker says, "I know this group is going to come up with some great ideas. Sam, I am going to ask you to be the recorder for the group. We will start with Liam and go around the group and end with you. While they share ideas, please write them down. Liam, when we share as a class, I would like you to be the reporter."

Liam responds saying, "Okay, I guess I would want to know how much space he has?"

Yancey adds, "How big will he make it?"

"What shape will he make it?" Talia adds.

Sam offers, "What will he make it out of?"

When it is Liam's turn again, he states, "I'm not sure what else he would need to think about. Someone else can go."

Yancey suggests, "What kind of sand will he put in it?"

Talia asks, "Where will he get the supplies?"

Sam adds, "And how will he keep animals out of it? Our neighbor's cat always went to the bathroom in our sandbox."

"I got one now," Liam interjects, "how much sand will he put in it?"

Sam asks, "What about how much will it cost?"

> Liam responds with enthusiasm, "Oh, that is a good one. Anything else?"
>
> After the round, students review their lists. Mr. Baker checks in with the groups. He says, "Great job, everyone. You've come up with a lot of ideas. This helps you see how many factors go into building something like a sandbox. Everyone had a chance to share, and this made it a real team effort." •

In summary, Mr. Baker supports **creative thinking** by allowing students time and space to brainstorm questions and to look at the solution to the sandbox problem from different perspectives. He promotes **adaptability** by encouraging students to look back at what they have done to incorporate new information. He promotes **social awareness** as he continually encourages students to work collaboratively, providing them tools and structures to make interactions more constructive. As peers experience confusion and frustration, Mr. Baker encourages students to show compassion and **empathy**.

Mrs. Vaughn and Her High School Math Class

After playing basketball in PE, Mrs. Vaughn hears a student complaining that he didn't play well because the ball was flat. Others begin talking about the "normal" bounce height of a basketball, and this led to the bounce height of other balls they commonly used, such as a volleyball, soccer ball, and tennis ball. Mrs. Vaughn leans into student curiosity and asks, "What type of ball is bounciest?" To answer this question, students will need to experiment with different types of balls and at the same time control variables. She recognizes that being able to make decisions effectively during the modeling process and when reporting the results is of utmost importance for the solution to adequately impact the real-world situation. She also recognized that social awareness, specifically group dynamics, would impact the effectiveness of the collaboration. The goal for math class became using mathematical modeling to describe the bounce height of a particular ball to compare the bounciness.

Over the semester, students practiced distinguishing between situations that can be modeled with linear functions and exponential functions (F.LE.1) and others. The following is an excerpt from a lesson that specifically supports students in recognizing situations in which one quantity grows or decays by a constant percent rate per unit interval relative to another (F.LE.1c). In addition, students are expected to learn to construct exponential functions given a table of data (F.LE.2).

Mrs. Vaughn begins by asking students to make assumptions and define the variables. She tells students they need to use both **curiosity** and **creativity** to list what they already knew and what they needed to know. They started with some questions: "What affects the bounce height?" "Does size matter?" "Does weight matter?" "Does the material used to make the ball matter?" For balls that hold air, does air pressure matter?" (Science Buddies, 2024)

Students researched factors affecting bounce height, discovering that air pressure played a significant role. They then focused on comparing the bounce heights of different balls and discussed methods in small groups. After sharing ideas, they decided each group would measure the bounce height of one ball. To control variables, the class agreed to ensure the ball's PSI was within normal limits, and use the same initial drop height, method, and data organization. They established a data collection procedure and table before beginning the investigation, which included basketballs, volleyballs, soccer balls, tennis balls, golf balls, and Ping-Pong balls.

> Mrs. Vaughn announces, "You've all been assigned to a group with a specific type of ball. Start by discussing each member's strengths and deciding who will do what. Make sure everyone contributes, and consider trading roles if needed."
>
> As Mrs. Vaughn circulates, she overhears Trey say, "We need someone to drop the ball and someone to record the bounces for measurement."
>
> Julie adds, "Someone should write the measurements in the table. What else do we need?"
>
> Tatiana suggests, "I think we should have two people video record so we can compare. I'll organize the data in the table."
>
> Trey responds, "I'll drop the ball. Can you two handle the recording?" He points to Julie and Zayne.
>
> After the data is collected, Mrs. Vaughn tells the class, "Take three minutes of quiet time to look for patterns in the data and decide what kind of

function best models it. Think about how you'll explain your ideas to the group. Set a timer for three minutes."

Once the timer goes off, the students begin discussing their findings. Mrs. Vaughn listens in as they share their thoughts.

She hears Kai say, "I saw that it is not linear."

"I agree," says Paulie, "the amount it goes down by varies so I think it is exponential."

Darius adds, "It could also be a power model. They can look like that too."

Kai replies, "But these seem to have a constant ratio, so I think it is exponential. We can assume that it is and do exponential regression to see if we have a good fit."

"How do you do that?" Joy asks and then says, "I don't remember."

Darius says, "I will show you how when I do it on my calculator so do what I do."

After each group records their data in a table and writes an exponential model on a piece of poster paper, they then hang it on the board and identify a group member to explain what the model described about their ball. After a discussion on which ball is bounciest, to encourage curiosity and creative thinking, Mrs. Vaughn asks students what this makes them wonder about now. For example, students ask, "What if the start height was changed? What if they were put in a freezer before bouncing? What about the time it takes to fall? Does weight make a difference?" Before ending class, the teacher asks each student to reflect on their work and list two ways they contributed to the group's work. ●

Students in this lesson were encouraged to be curious and creative. They had many opportunities to notice and wonder, to generate questions, to investigate, and to create a way to share. Mrs. Vaughn also highlighted for students the many decisions they needed to make. She encouraged them to work in groups to do this and provided collaborative structures to hold students accountable for contributing, which supports the development of social awareness.

Reflection

Across both the narratives, the teachers provided time for students to define the problem, identify variables, make assumptions, and determine what to include and what information is needed. Students generated questions and presented them to peers. The teacher then focused the math lesson to address a specific content standard. At this point, students had a good understanding of the real-world model and were ready to translate into a mathematical model. Looking at students' individual and small group work, as well as listening to group discussions and responses to class discussions, enabled the teachers to have a general sense of how well students are understanding the content and engaging in modeling practices. They also listen for and observe decision-making processes, and provide structures and interventions as needed.

SUMMARY

In this chapter we examined how we can leverage MP4 and draw on the social-emotional competencies of **responsible decision-making**, **social awareness**, **creative thinking**, and **adaptability** to enhance classroom lessons. The goal of MP4 is to support students as they solve real-world problems and model situations. We shared how modeling with mathematics is more than solving word problems. Modeling problems are messier and require students to ask questions, seek information, and interpret the results. We demonstrated how you can integrate intrapersonal and interpersonal skills into math lessons that involve MP4.

All the competency builders we shared in this chapter provided ways to explicitly address intrapersonal and interpersonal skills. The classroom narratives illustrated what this would look and sound like. When regularly incorporated into math lessons, this approach seamlessly becomes part of the learning experience, strengthening concepts and promoting both individual and collective reflection. Integrating social-emotional competencies with MP4 helps students become more effective problem solvers, critical thinkers, and lifelong learners in mathematics and beyond.

It is one way as math teachers we can develop a positive mindset and emphasize for students the importance of clearly defining and understanding a problem, whether it is a mathematical problem or everyday problem, before attempting to solve it. Now you must make thoughtful and purposeful pedagogical choices, employing a variety of instructional tools and methods to make this come alive for students.

Questions to Think About

1. How and when do I provide students with real-world problems that require them to create and apply mathematical models?

2. In what ways do I encourage students to listen to and consider the perspectives of others when discussing models and solutions?

3. How do I create an environment where students feel safe and supported when sharing their models? How do I encourage students to provide constructive feedback on each other's models?

4. What prompts do I use to encourage students to reflect on their modeling process and the decisions they made while constructing their models?

5. How do students respond when I encourage them to adapt their models when they receive new information or feedback? How can I support this?

Actions to Take

1. Find a traditional word problem in your textbook and give it a makeover!
2. Explore Three-Act Math Tasks.

CHAPTER 5

USING MATHEMATICAL TOOLS TO BUILD ADAPTABILITY AND DECISION-MAKING SKILLS

You may have heard the question, "Is it the right tool for the job?" It is a question that can be applied in different ways depending on the context in which it is used; but it is asking, when you have a job to do or problem to tackle, what tool will produce the best results? J. SanGiovanni (2015) describes a mathematical tool as "anything that aids in accomplishing a task." A tool could be a physical object like a ruler, compass, or manipulative. It could be a pictorial or symbolic representation like a table, graph, diagram, or graphic organizer. Tools may include technology, such as a graphing calculator, computer program, or a mathematics app. However, tools also include the prior knowledge, skills, and competencies students bring with them to the mathematics classroom and leverage to solve problems and reason mathematically. Mathematical Practice 5 (MP5): Use appropriate tools strategically, empowers students to choose and use the most efficient tool with guidance from the teacher.

> **USE APPROPRIATE TOOLS STRATEGICALLY**
>
> *Mathematically proficient students consider the available tools when solving a mathematical problem. These tools might include pencil and paper, concrete models, a ruler, a protractor, a calculator, a spreadsheet, a computer algebra system, a statistical package, or dynamic geometry software. Proficient students are sufficiently familiar with tools appropriate for their grade or course to make sound decisions about when each of these tools might be helpful, recognizing both the insight to be gained and their limitations. For example, mathematically proficient*
>
> *(Continued)*

> (Continued)
>
> *high school students analyze graphs of functions and solutions generated using a graphing calculator. They detect possible errors by strategically using estimation and other mathematical knowledge. When making mathematical models, they know that technology can enable them to visualize the results of varying assumptions, explore consequences, and compare predictions with data. Mathematically proficient students at various grade levels are able to identify relevant external mathematical resources, such as digital content located on a website, and use them to pose or solve problems. They are able to use technological tools to explore and deepen their understanding of concepts. (CCSSM, 2010)*

When understanding MP5, we must recognize there are two processes taking place—choosing and using. Students must attend to these processes both appropriately and strategically. It is important to not limit this practice to focus only on the "appropriate" choice or use of the tools. In and of itself, using a mathematical tool appropriately does not constitute full engagement with MP5 (Matney et al., 2020). Table 5.1 illustrates actions students can take as they engage with MP5, specifically as they decide how to choose and use tools in the mathematics classroom appropriately and strategically.

Table 5.1 • *MP5 Student Actions*

	Appropriately (Doing something suitable or acceptable under the circumstances)	**Strategically** (Makes best use of available resources to complete a task)
Choose	Selects a reasonable tool to complete a task or solve a problem by asking, • What tools do I have access to? • What tool makes sense to use with this task/problem?	Intentionally chooses a tool that allows them to efficiently solve the problem by asking, • What tool would allow me to complete the task or solve the problem most efficiently?

Use	Uses the tool correctly and for its intended use to get accurate results by asking, • Does my use of this tool fit with its purpose? Uses the tool with integrity by asking, • Am I using the tool in an honest and truthful way?	Checks for reasonableness throughout by asking, • Is this tool helping me reach my goal or do I need to select a different tool? • Was there a tool that would have been more efficient?

 Download this table at https://companion.corwin.com/courses/wellroundedmathstudent

In elementary grades, students *choose* counters or cubes to help them solve an addition or subtraction problem and *use* them appropriately and strategically to count on or count back. They may *choose* a ruler to measure an item and *use* it appropriately by lining the edge of the ruler with the edge of the item. In secondary classrooms, students might *choose* paper and pencil to graph a function or, for more complex functions, might find it more appropriate to *choose* technology software and *use* it to input data to create a graph. Within the ever-changing landscape of technology, one might immediately gravitate toward the technological tools that are available for students; however, it is important to also consider how using concrete manipulatives and other tools can provide students with the experiences necessary to develop intrapersonal and interpersonal skills through their engagement with MP5.

MP5 provides opportunity for students to practice self-regulation and promotes **integrity**, **decision-making**, and **adaptability** as students choose and use tools both appropriately and strategically to solve problems involving mathematics. A key insight within this mathematical practice is developing the skill set to identify which tool is not only appropriate, but efficient for the specific problem they are attempting to solve. Students will grapple with various tools, evaluate the efficiency of each tool, and effectively use tools to solve problems (Matney et al., 2020). These experiences allow students to prepare for career and life, extending beyond their PreK–12 academic success.

MERGING CONTENT STANDARDS, MATHEMATICAL PRACTICES, AND SOCIAL-EMOTIONAL COMPETENCIES

Remember that planning and preparing mathematics lessons that explicitly incorporate MP5 and intrapersonal and interpersonal skills includes multiple decision points. Thinking back to the framework introduced in this book, start connecting the mathematical goals and practices, and intentionally bringing out the social-emotional skills.

Mathematical Content Standard and Corresponding Mathematics Goal

Begin by asking, **"What is the mathematics goal of this lesson?"** Consider the second-grade standard in which students are asked to "Measure the length of an object by selecting and using appropriate tools such as rulers, yardsticks, meter sticks, and measuring tapes" (2.MD.A.1, CCSSM, 2010). In this standard, the language explicitly showcases the selection process and appropriate use of various tools. One goal for a lesson addressing this standard would focus on identifying which tool would be most efficient for measuring the length of a particular object (and why) and then using the tool appropriately. It may be tempting to tell the student what tool to use, show them how to use it, and then have students practice using the tool. However, this only addresses part of the standards, in that students demonstrate they are proficient with the skill of measuring with a given tool, but this gives us no insight into their understanding of when to choose a particular tool. Students must explore various tools to deeply understand their purpose. Appropriate and efficient use extends beyond the notion of just using each of the tools to measure, but also critically examining each tool for benefits and limitations based on the specific context.

At the secondary level, consider the standard "Use coordinates to compute perimeters of polygons and areas of triangles and rectangles, e.g., using the distance formula." (HSG-GPE.B.7, CCSSM, 2010). This standard focuses on students practicing with the distance formula and understanding its connection with the Pythagorean theorem. Students use mathematical formulas and properties as well as graphic representations as tools to solve problems. They do this by using the coordinates of the vertices of triangles, rectangles, and polygons graphed in the coordinate plane to find perimeter and area. They use other formulas as well. For example, they use the coordinates and apply the slope formula to determine whether segments are perpendicular so they can find perimeter and area. A goal for a lesson aligned to this standard would be for students to choose and use the appropriate properties and formulas to solve problems involving perimeter and area on a coordinate grid.

Mathematical Practice

After unpacking the mathematics from the content standards, next ask **"Which mathematical practice supports engagement in this content standard?"** Some content standards use the word *tools*, which makes a direct connection to MP5. However, many other standards that align to MP5 are not as explicit. In this case, we notice terms like *consider*, *analyze*, *explore*, and *compare* along with language stating students should perform a mathematics-related task naming specific tools to be used (e.g., manipulatives, drawings, tables, graphs, coordinates, equations, models, calculators, spreadsheets, dynamic geometry software, and graphing technology). Tools are valuable in helping students visualize the concepts they are studying. Additional examples of standards that have an MP5 focus include the following:

- Use multiplication and division within 100 to solve word problems in situations involving equal groups, arrays, and measurement quantities, e.g., by using drawings and equations with a symbol for the unknown number to represent the problem. (3.OA.A.3., CCSSM, 2010)

- Construct a function to model a linear relationship between two quantities. Determine the rate of change and initial value of the function from a description of a relationship or from two (x, y) values, including reading these from a table or from a graph. Interpret the rate of change and initial value of a linear function in terms of the situation it models, and in terms of its graph or a table of values. (8.F.B.4., CCSSM, 2010)

The language within these content standards states that students will use a variety of appropriate mathematical tools to solve developmentally appropriate mathematical problems. MP5 focuses on both the choice of tools and use of strategic decisions based on efficiency and appropriateness. Students must use a variety of social-emotional competencies to examine tools for efficient problem-solving.

Social-Emotional Competencies

Next ask, **"What interpersonal and intrapersonal skills are inherent, are needed, or can be further developed for students while engaging in MP5?"** These skills can be identified by examining the action verbs in MP5 and highlighting the underlying skills necessary to engage in these actions. To be proficient with MP5, students should "consider the available tools when solving a mathematical problem . . . make sound decisions about when each of these tools might be helpful . . . detect possible errors by strategically using estimation and other mathematical knowledge . . . explore consequences, and compare predictions" (CCSSM, 2010).

The integration of intrapersonal and interpersonal skills with MP5 requires that we are mindful of these connections and purposeful in their implementation. In this chapter, integrity, responsible decision, and adaptability are the targeted competencies.

Intrapersonal Skills

While the description of MP5 doesn't specifically state that mathematical tools are to be used with **integrity**, it is clearly implied. The idea that tools are always used appropriately is a misplaced assumption, as students find many ways to utilize technology and tools. Clear guidelines and expectations must be provided to promote academic honesty related to tools. Students must continue to discuss and revisit the importance of **integrity** and **responsible decision-making** so they see the importance of using the tools strategically and appropriately. Students must be given the opportunity to use a variety of tools to solve a problem and develop a viewpoint about the use of those tools for a specific context. During this decision-making process, students must consider the available tools and their efficiency. Students develop and use **integrity** in the exploration phase of selecting and using tools as they must **make responsible decisions** about "when each of these tools might be helpful" in a specific situation (Table 5.1).

Interpersonal Skills

Students leverage and build **adaptability** as they strategically use tools to determine efficiency. As students "detect possible errors, explore consequences and compare predictions," they are allowed the opportunity to adjust the mathematical model, problem-solving process, or solution. Setting clear expectations regarding this fluid trial-and-error type of process is a vital step to promote perseverance during this development of adaptability.

Instructional Structures and Engagement Strategies

Next, we shift toward planning and ask, **"With an eye on our mathematics goal, how will I support social-emotional development as I engage** learners in MP5: Use appropriate tools strategically? What structures, strategies, methods, and/or tools can I use?"

The appropriate and strategic use of tools is dependent on students using the tools with integrity, so it makes sense to start by addressing **integrity** explicitly in the mathematics classroom. Using a tool with integrity must be explicitly emphasized. There are just as many competency builders in here for teachers as there are for students because if we aren't planning to use the technology appropriately, it's very hard for students to learn to use it appropriately!

Teaching Students What Integrity Means

The International Center for Academic Integrity (ICAI, 2021) encourages educators to promote the development of integrity in learners of all ages based on "six fundamental values: honesty, trust, fairness, respect, responsibility, and courage." In the mathematics classroom, these values should be discussed in a variety of ways specific to MP5, with a focus on using tools to expand our learning, not "cheat" in a traditional sense. It is about doing the right thing when the teacher is not watching. Consider the use of calculators. If the lesson aims to teach a computational concept, using a calculator may undermine the integrity of the learning process, as it can diminish both conceptual and procedural understanding. However, if the goal of the lesson extends beyond the computational concept, using a calculator can enhance and streamline activities, creating a richer learning experience.

Competency Builder 5.1
Fostering Core Values Associated With Integrity

Integrity development can be infused into the daily classroom culture by explicitly teaching about and using the vocabulary that reflects the core values identified in Table 5.2 (Price-Mitchell, 2015). Begin by asking students to work with a partner to define the values and describe what they look and sound like in a mathematics class. Engage in a class discussion where students share their ideas and create a version of this to post in the classroom. Use these words during lessons to encourage students to identify when they are relying on these values and name the specific value to further develop these skills.

(Continued)

(Continued)

Table 5.2 • *Core Values Anchor Chart*

Fundamental values of academic honesty	Definition	What it looks like and sounds like in mathematics class
Responsibility		
Respect		
Fairness		
Trustworthiness		
Honesty		
Courage		

online resources: Download this table at https://companion.corwin.com/courses/wellroundedmathstudent

Summarize by discussing the importance of academic integrity. Share with your students that this means taking responsibility for their work, holding themselves and others accountable for using tools ethically, understanding the consequences of misusing tools to cheat, and helping students understand that these things build a reputation of trustworthiness.

Planning With Intention

To engage learners in using tools to support the development of their mathematics content knowledge, be intentional in planning to use tools in a way that fosters their technical skills and their ethical understanding. The SAMR model for technology integration outlines four classifications of integration, including substitution, augmentation, modification, and redefinition, and can be utilized as you consider how to engage students with tools (Puentedura, 2014). Each degree of the model intensifies from basic use of the tool in place of another (or no tool) to using tools to create or complete tasks that would otherwise not be possible (Romrell et al., 2014). As you plan your lessons, consider whether you want to transform the learning opportunity by replacing mental mathematics or paper and pencil with another

CHAPTER 5 • USING MATHEMATICAL TOOLS TO BUILD ADAPTABILITY

tool or if you are hoping to transform the learning opportunity by engaging students in strategic use of tools that allows them to explore situations and uncover mathematical concepts that may otherwise be unattainable.

This transformation level of the planning process starts with task selection. For MP5, we must select tasks that

- Allow our students to grapple with the exploration of a variety of tools.
- Promote collaborations with their peers.
- Are open enough to encourage students to try using tools of their choice.
- Allow students to make decisions about these tools and justify their thinking.
- Identify the benefits and limitations of tools to determine efficiency.
- Encourage students to think deeply about the purpose of the tools they use and how they are used appropriately and strategically.

Teacher planning prompts support a process for effectively implementing a task to align to MP5.

Competency Builder 5.2
Planning Questions

Once you have selected an appropriate task, use the guide provided in Table 5.3 to help you plan and implement the purposeful and explicit use of tools in a way that fosters the development of integrity. Consider each prompt and record your own response in the second column. Take the necessary steps to integrate your responses into your lesson plan.

Table 5.3 • *Planning Questions With an Eye on Integrity*

Planning Prompts	Response
What tools do we have in the classroom that students can use to support them in accessing the mathematics content and how will it be used? Make sure students know what tools are allowed to ensure fairness.	

(Continued)

THE WELL-ROUNDED MATH STUDENT

(Continued)

How will I introduce them to the tool and make clear the purpose? Establish clear guidelines for when and how tools can be used to ensure they are aligned with promoting fairness and honesty.	
What expectations do I have for students as they use these tools? What does honest and trustworthy use of tools look and sound like?	
What directions will I give and/or what procedures will I ask them to follow as they use the materials? Be clear about responsible and respectful use of the tools.	
How do I anticipate students will engage with the materials (e.g., individually, pairs, small group, whole class)?	
How will I have my students reflect on their appropriate and strategic use of the tool (e.g., discuss with a partner, share with the class, written response to a prompt, self-reflection survey, etc.)?	
How will I collect feedback regarding the effectiveness of the tool?	
How will I collect evidence regarding the development in integrity?	

online resources Download this table at https://companion.corwin.com/courses/wellroundedmathstudent

Integrating the use of tools with building integrity can help students become both proficient users of mathematics tools and individuals capable of making fair, honest, responsible decisions about how to apply these tools in ethical ways.

Giving a Textbook Task a Makeover

Closely examine your instructional resources to evaluate how well the problems actively engage your students in tasks that allow for strategic and appropriate use of tools. It is not uncommon to find textbook activities where students are measuring pictures in a workbook or step-by-step instructions on how to use a specific tool. Although these are important parts of the learning process, these activities need to take place within the context of an exploration that involves decision-making and selection of tools for specific situations. For example, when choosing a ruler as the best measuring tool to use, this would be appropriate for smaller items, but if measuring a wall, it might not be the most efficient. Tasks designed to encourage students to gain this insight would allow them to try measuring the wall with a ruler and thus recognize that this might not be the most efficient tool. As the facilitator, you will ask students questions so they can reflect on their actions, evaluate their choices, and consider other alternatives.

Competency Builder 5.3
Comparing Tasks to Encourage Appropriate and Strategic Use of Tools

Consider the two tasks that follow as you make instructional decisions related to tasks that will engage your students in MP5.

Task A:

In the following image, find the length of each object to the nearest unit of measure.

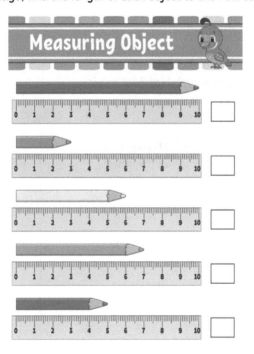

Source: istock.com/rightholder

(Continued)

(Continued)

Task B:

Provide students access to various tools, including a ruler with inches and centimeters, a tape measure, a meter stick, etc. Explain to students that they will work with a partner to measure and record the length of each item, deciding which tool to use. Tell students to each perform their own measuring and recording and then compare the results. If they get different results, they should discuss why they are different and agree on a measurement. To support appropriate use of the tools, introduce (or review) measurement processes by asking students what they must do when measuring to get an accurate measurement and then demonstrating it. This includes:

- Identify where to start and where to end. Make sure the tool lines up at the start and the measurement is marked at the end.
- Measure in a straight line.
- There should be no spaces or overlap between measurements if a tool is used repeatedly to find a length.

Direct students to record their measurement, unit of measure, and the tool used. Remind them to compare with their partner and be prepared to explain why they chose the tool they used.

Table 5.4 • *Measurement Mania Activity*

Measurement Mania		
Item	Measurement (with label)	Tool Used
Width of the classroom door		
Length of a paperclip		
Length of a box of tissue		

Height of the table		
Width of a book		
Length of the whiteboard		
Width of a pencil		

Download this table at https://companion.corwin.com/courses/wellrounded mathstudent

Consider both tasks. In what ways are students engaged in measuring and using tools? Which task would enable students to demonstrate their measuring and decision-making skills? Take note, the pencil is being measured in an unlabeled unit; the work of deciding which tool to use, how to use the tool, and determining the unit of measure is all done for students. This lowers the cognitive demand and opportunity to choose and use a tool strategically and appropriately. Task B provides relevancy, as students are measuring the actual items, not images of the items. Students make and justify decisions about what tool to use, what unit of measure to use, and how to make the measurements. It also encourages adaptability, as they may want to use a different tool than a peer and they can discuss why using centimeters instead of inches on a ruler would be more accurate for smaller items, but not necessary for larger items. When implementing a task like this, you need to explicitly address decision-making and adaptability with students by naming the skills and asking students to reflect on these competencies. The task can be adapted depending on the grade level, skill level, and measurement concepts.

Setting Guidelines for the Use of Tools

Success using tools in the classroom requires teachers to be intentional and explicit about how they are to be used.

> **Competency Builder 5.4**
> *Implementing Guidelines for the Use of Tools*
>
> We can leverage "7 Musts for Using Manipulatives" (Burns, 1996) and apply them to tools in general. The following actions support effective use of tools in the classroom. Discuss these guidelines with colleagues and take appropriate actions to implement them in the classroom:
>
> 1. Discuss with students how tools can help them learn mathematics and be clear about the mathematics concept they will learn using them.
> 2. Set rules for using the tool. Talk about how using manipulatives to learn in class is different from playing with a toy.
> 3. Decide how and where to store tools for students to easily access.
> 4. Determine how students will familiarize themselves with the tools, including exploring the tools without a mathematics focus.
> 5. Create written reminders for students to visually reference about the tools they can use and include their ideas about the tools after they have experience with them.
> 6. Have students' journal or free write about their experiences with tools. Specifically ask them to reflect on the choices they made and instances where they changed tools.
> 7. Encourage parents to engage with the tools and ask their students about their experiences.
>
> These actions highlight the need to involve students in the implementation of tools. Encourage students to ask for help if they are unsure how to use the tools appropriately; this fosters courage and builds trust. Encourage students to ask why certain tools work and to explore the mathematics behind them. A metacognitive approach to the use of the tools will help students see that tools are not shortcuts to answers but instead provide opportunities for deeper understanding.

Using Questioning to Engage Learners

Questioning during inquiry-based or student-centered learning opportunities allows the teacher to assess and advance student thinking without lowering the cognitive demand of the task (National Council of Teachers of Mathematics [NCTM], 2014, 2017). Questions can be designed to specifically focus on the use of tools.

CHAPTER 5 • USING MATHEMATICAL TOOLS TO BUILD ADAPTABILITY

Competency Builder 5.5
Questions to Support Appropriate and Strategic Use of Tools

Teach students to think critically about the choice of the tool and about the results provided by the tool. Help students to see the goal is not just getting the "right" answer but understanding the process. Questions that support appropriate and strategic use of tools emphasize (1) student exploration of a variety of tools, (2) student choice regarding which tool is most appropriate for the specific mathematical situation, and (3) an underlying focus on student justification of their choices and reasonableness of the results. These three foci promote the development of adaptability, responsible decision-making, and integrity. Use the following questions (Kansas State Department of Education, 2017, p. 13) and support students in learning how to ask these questions of themselves:

- What mathematical tools could we use to visualize and represent the situation?
- What information do you have?
- What do you know that is not stated in the problem?
- What approach are you considering trying first?
- What estimate did you make for the solution?
- In this situation would it be helpful to use a graph? Number line? Ruler? Diagram? Calculator? Manipulative? Why?
- Was it helpful to use . . . ?
- In what situations might it be more informative or helpful to use . . . ?

Focusing on the Benefits and Limitations for Tools

Students need to be familiar with a variety of tools and know how to use and when to choose a particular tool. MP5 states students should be able to recognize the insight gained and their limitations.

Competency Builder 5.6
Using Graphic Organizers for Tool Selection Evaluation

Use an organizer to chart the use of tools and their benefits and limitations (Table 5.5). As a new tool is introduced, put the name of the tool in the first column. After using the tool, add a description of the benefits and limitations of the tool.

(Continued)

(Continued)

Table 5.5 • *Example of an Organizer for Selecting a Tool: Benefits and Limitations*

Tool	Benefits	Limitations
Paper/pencil		
Calculator		
Graphing calculator		
Excel spreadsheet		
Photomath		
Artificial intelligence		

online resources Download this table at https://companion.corwin.com/courses/wellroundedmathstudent

Revisit the chart each time a tool is used and briefly discuss the appropriate use of the tool based on the benefits and limitations. Continue to add to the benefits and limitations as needed. Talk with students about how this information helps them in making decisions about what tool to use and how a similar type of chart can help in other decision-making situations in life. Rather than avoiding tools such as Photo mathematics and other AI tools, recognize that these tools are not going away and are constantly being developed. Be intentional by asking how these tools can be used to appropriately and strategically support mathematics engagement and deep understanding. In this way, we support the development of integrity.

Assessing Interpersonal and Intrapersonal Skills

Next ask, "How will I assess students' progress toward the mathematics goal of this lesson, their engagement in the mathematical practice standard, and their ability to use and continue to develop 21st-century skills? How will I provide feedback? How will I build an opportunity for students to reflect on the development of their 21st-century skills?"

CHAPTER 5 • USING MATHEMATICAL TOOLS TO BUILD ADAPTABILITY

Using formative and summative measures both formally and informally alleviates the pressure often felt by teachers, and students, to collect evidence of learning (Burke, 2010).

Consider again the classroom measurement example in Competency Builder 5.3. Assessment like this task would involve observing the appropriate use of the tool (lining up the edge of the tool with the edge of the item), the accuracy of the measurement, the appropriate use of units, and the skill of measuring multiple items using a tool. You can observe their progress toward the mathematics goal, the appropriate and strategic use of the tools, and the **integrity**, **adaptability**, and **decision-making** displayed by students. For example, use checklists or observation tools like those in Tables 5.6, 5.7, and 5.8, to informally assess students' engagement in MP5 and development of social-emotional competencies. Offer the following prompts during or after a lesson to help students reflect on their own learning.

- Look at the Core Values Anchor Chart (Table 5.2). How did you demonstrate **integrity** today? Provide at least one example.
- As you used various tools, how did you **adapt** your thinking or actions? Provide at least one example.
- What needs to be considered when making decisions about the tools you use to help you understand and solve problems? How do they help with decision-making?
- How do the tools you use help you understand mathematics more deeply?

Table 5.6 • *Whole Class Observation Tool*

Name	Mathematics Goal	Engagement in Practice Standard *Chooses appropriate tool*	Engagement in Practice Standard *Uses tools strategically*	Intrapersonal Competency *Integrity*	Interpersonal Competency *Adaptability*

(Continued)

(Continued)

Name	Mathematics Goal	Engagement in Practice Standard *Chooses appropriate tool*	Engagement in Practice Standard *Uses tools strategically*	Intrapersonal Competency *Integrity*	Interpersonal Competency *Adaptability*

Note: *Progress will be marked using 0–No evidence, 1–Little evidence, 2–Adequate evidence*

Download this table at https://companion.corwin.com/courses/wellroundedmathstudent

Table 5.7 • *Individual Observation Tool*

Name of student:

Mathematics Goal	No Evidence	Little Evidence	Adequate Evidence
Engagement in the Practice	No Evidence	Little Evidence	Adequate Evidence
Chooses appropriate tool			
Uses tool strategically			
Social-Emotional Competencies	No Evidence	Little Evidence	Adequate Evidence
Integrity			
Responsible decision-making			
Adaptability			

Download this table at https://companion.corwin.com/courses/wellroundedmathstudent

Table 5.8 • *Self-Assessment Checklist*

Social-Emotional Competency	Not Sure	Not Yet	Getting There	Got It
Integrity				
Decision-making				
Adaptability				
Other skills used:				

Download this table at https://companion.corwin.com/courses/wellroundedmathstudent

LOOKING AT EXEMPLARS IN ACTION

Now that we have explored the merging of content standards, mathematical practice standards, and social-emotional competencies, let's look more closely at an elementary and a secondary example, using the standards and competencies focused on in this chapter.

Mrs. Byrns and Her Second-Grade Class

Mrs. Byrns's second-grade class is focused on measuring the length of an object by selecting and using appropriate tools, such as rulers, yardsticks, meter sticks, and measuring tapes. She has selected MP5: Use appropriate tools strategically, as the practice standard focus (2.MD.A.1., CCSSM, 2010). As you read the classroom vignette, consider how Mrs. Byrns involves students in the development of their integrity, decision-making skills, and adaptability as they engage in MP5. Identify how this could look in your classroom.

> Mrs. Byrns starts the class with an open discussion. "The school wants to replace the old baseboards. We need to figure out how much material to order. What tools can we use to measure the baseboards on the walls?" After sharing their ideas with the whole class, Mrs. Byrns states, "You all said that you needed to use a tool, but you all chose something different to

(Continued)

(Continued)

use. Today, we will explore which tool is best, or most efficient, for this task. You will use the tool you chose, then share your process with the class."

Mrs. Byrns then lays out different tools students can use as they solve this problem, such as rulers, metersticks/yardsticks, tape measures, and a stack of recycled paper (after hearing a group's ideas). She says, "Now it's time to pick your tool. You will have ten minutes to measure the walls. I won't be able to be with every group at the same time so you will have to use your tools and perform the measurements with integrity. That's an important word so let's talk about what that means." She provides a student-friendly definition of integrity, telling the students it means they will do the right thing even when they are not being watched. They talk about what that would look like and what it would not look like.

She monitors each group, using a whole class observation tool similar to the one provided in Table 5.6. Her observations are displayed in Table 5.9.

Table 5.9 • *Mrs. Byrns's Whole Class Observation Tool*

Name	Engagement in Practice Standard	Engagement in Practice Standard	Intrapersonal Competency	Intrapersonal Competency	Intrapersonal Competency
	Chooses appropriate tool	Uses tools strategically	Responsible decision-making	Integrity	Adaptability
Group 1	1	1		1	2
Notes: Wanted to remove trim, redirected by asking them to reconsider and explore the options of tools.					
Group 2	1	1	1	1	2
Notes: Changed from ruler to meterstick.					
Group 3	1	2	1	1	1
Notes: Measured somewhat accurately with meterstick.					

CHAPTER 5 • USING MATHEMATICAL TOOLS TO BUILD ADAPTABILITY

Name	Engagement in Practice Standard	Engagement in Practice Standard	Intrapersonal Competency	Intrapersonal Competency	Intrapersonal Competency
Group 4			1	1	
Notes: Measured somewhat accurately with tallest student in the group.					
Group 5			1	1	
Notes: Measured somewhat accurately by laying pieces of paper end to end.					

When she asks, "Did everyone get the same measurement?" the class responds, "No!" Mrs. Byrns prompts, "Talk to your partners about why." Mrs. Byrns listens to student discussions and after one minute she strategically calls on several students to share their ideas. Then she says, "This time, discuss what we could do differently, so we get the same or at least similar measurements." Mrs. Byrns continues using the whole class observation tool during these partner discussions, noting students who were using language around the need for a tool that was helpful at measuring long lengths.

To help students focus on the tools they chose, Mrs. Byrns has the class create and complete the table of benefits and limitations for each tool, like Table 5.5.

Mrs. Byrns then leads a discussion on the tools used. "What are the benefits of using a person to measure?" Talia responds, "You always have yourself!"

"What about limitations?" Mrs. Byrns probes. Jonas replies, "It was hard because he had to move, which led to errors."

Kylie from Group 5 chimes in saying, "Same with us. When we set the papers out, some of them overlapped a little or there was space between, and we tried to keep them from moving, but it didn't always work."

Mrs. Byrns clarifies this saying, "So, let me restate that and you tell me if I got it. Using papers and the person's body were limited because they had to be moved and that led to errors in measuring. Is that accurate?"

(Continued)

(Continued)

"Yes," replies Antonio. "We just used regular paper (copy paper) but it would be different if we used smaller or bigger."

"Good point," Mrs. Bryns exclaims. "I think we might have found another limitation! Paper, and the human body, are all different sizes, so these are called non-traditional units of measurement. Can you think of any others?" Students chime in with other examples.

Mrs. Byrns summarizes saying, "Yes, these are all used regularly but are not always the same size. So that's a limitation of all non-traditional units of measurement. You also had to use it over and over again, which led to some mistakes."

Runchu adds, "That was the same for the meterstick. We kept having to pick it up and hold our finger in the spot, but sometimes we moved a little, so I don't think ours is going to be perfect."

"Let's compare the tools a little more," prompts Mrs. Bryns. "Would there ever be a time when using a ruler or tape measure might be more efficient, meaning it would work better and easier? I want you all to think hard about this so talk with your shoulder partner." After observing their discussions, Mrs. Byrns points out, "That type of thinking is what makes you a good decision-maker. To be a good decision-maker, we have to be able to identify our options and pick the one that is best in that situation. Not all situations are the same. It is like that in life too!" ●

The vignette demonstrates how Mrs. Bryns was able to encourage the appropriate use of a variety of tools in one lesson, while promoting intrapersonal and interpersonal skill development. Mrs. Byrns supported students' adaptability by discussing when one measuring tool may be more efficient or strategic than another tool; however, the focus on using standard over nonstandard measuring tools was supported by student experiences. She promoted student integrity by allowing them time to work on their own and try using the tools, while monitoring and encouraging them to be honest in their work and discussion of the process, even if others had different

CHAPTER 5 • USING MATHEMATICAL TOOLS TO BUILD ADAPTABILITY

ideas. This specifically came up in the class when groups wanted to change their strategy or tool. Questions like, "Why would you change?" helped promote the notion that when engaged in responsible decision-making, with integrity, one must consider multiple ideas and base their decision on the unique situation.

Mr. Heinen and His High School Geometry Class

Mr. Heinen's geometry class is studying coordinate geometry. Most recently, they have derived the distance formula using the Pythagorean theorem in the coordinate plane. Now, they are using that formula to calculate attributes (such as side lengths and diagonals) to classify quadrilaterals. The lesson focuses on the content standard, "use coordinates to compute perimeters of polygons and areas of triangles and rectangles, e.g., using the distance formula" (HSG-GPE.B.7, CCSSM, 2010). As you read the classroom vignette, consider how students are engaged in using tools appropriately and strategically, with integrity. The lesson began with an overview of the meaning of integrity, and students generated examples and nonexamples. This narrative is an excerpt from a lesson where integrity was in question.

As Mr. Heinen circulates the room, he notices Kennedy and Elliot frantically writing down something from a tablet. He sees a mathematics camera solver app is open. When they see him approach, Kennedy and Elliot look slowly from the screen to Mr. Heinen. Mr. Heinen asks, "How's it going?"

Kennedy points to the device and quickly explains, "We're using this mathematics tool to find the lengths of the sides and diagonals in our quadrilaterals."

Mr. Heinen nods. "Tell me, how are you using this tool?" he asks.

The students demonstrate how to take a picture of the mathematics problem, and it gives an exact answer and approximation. "Very resourceful," Mr. Heinen comments. He asks, "What are the benefits and limitations of this tool?"

Kennedy shares that it helps get the correct answer. Elliot mentions that using the app is easier than a calculator.

(Continued)

THE WELL-ROUNDED MATH STUDENT

(Continued)

Mr. Heinen nods. He says, "So this tool helps you find the answer, but does it help you understand how to perform the calculation?" Kennedy and Elliot are silent. He points to a special button on the screen. "Tell me—what does 'this' do?" Elliot and Kennedy push the button and it expands, displaying each step for the calculation. Kennedy explains that this is what they are writing in order to show their work. "I see," Mr. Heinen nods again. He asks, "Does it explain how these steps work?"

Elliot shakes his head no.

Mr. Heinen says to Kennedy and Elliot, "I appreciate that you thought about this tool to help with this task. However, the way you use the tool matters. If you use this tool to simply write down the steps and the answers without trying to understand the mathematics, that is not an appropriate use of it. It is actually a form of cheating. It makes you focus on the answer instead of on the learning."

The students' posture dips in embarrassment. "But," Mr. Heinen continues, "if you use this tool to understand how to solve problems using the distance formula, then it is not only appropriate it is also strategic. If you are stuck and use this tool as a resource, then you can learn from it. That is how you can use it with integrity."

"So, we can use this to understand how to calculate the distance formula?" Kennedy asks.

Mr. Heinen nods, making one more request. "If you chose to use this tool to calculate the distance between two points, you must explain 'why' the calculation happens for each step on your paper." They begin looking at the steps displayed on the tablet, talking through each step, and making sense of the mathematics.

As Kennedy and Elliot work, they notice something different from their class notes. They ask Mr. Heinen for his assistance. "In class," Kennedy begins, "we came up with this formula." She points to the equation:

$d = \sqrt{(x_1-x_2)^2+(y_1-y_2)^2}$. "But the program uses this formula." She points to the equation: $d = \sqrt{(x_2-x_1)^2+(y_2-y_1)^2}$. "I mean, they look kind of the same, but they are different. Which one is right?"

"Ah," Mr. Heinen realizes. "That is a really good observation. I mentioned this briefly in class, but it hasn't come up in our work yet. That is both a benefit and a limitation for using this tool. We are going to make sense of the equation you see so that is a benefit."

He gathers the class to discuss Kennedy and Elliot's findings. Drawing attention to the two equations, he recalls their earlier discussion and prompts students with questions to explore the similarities and differences between the formulas. As they analyze the equations with a partner, students discover that although the formulas look slightly different, both are valid. They then use the properties to generate other formulas that would work as well.

As the discussion ends, Mr. Heinen addresses the class. "When using tools like this, it's important to use them with integrity. Simply copying steps and answers without understanding the problem is not learning—it's cheating. When used properly, resources can help us learn, like calculating the distance between two points. There will be times when you encounter conflicting information, like we did today. In those cases, take the time to understand the problem, adapt your thinking, and figure out what's true. That's how you use tools strategically." ●

Mr. Heinen could have scolded Kennedy and Elliot for cheating. Instead, he seized an opportunity to discuss integrity and responsible decision-making, connecting it to the responsible use of resources. He connected the need to adapt and evaluate when something contradicts, or disputes, known information.

Reflection

To develop intrapersonal and interpersonal skills in math, teachers plan learning opportunities that balance content standards, practices, and social-emotional competencies. In the examples shared, teachers selected tasks, varied engagement strategies, and assessed students in ways that encouraged integrity, adaptability, and responsible decision-making. These approaches help students apply these skills to new situations by asking, "Can a mathematical tool help solve this problem?" Teachers guide this exploration with a focus on both mathematical concepts and social-emotional growth. These skills are not learned alone but embedded in the mathematics teaching framework. At the end of learning opportunities with alignment to MP5, teachers and students should be able to answer questions like "What mathematics did we learn? When can we use this tool efficiently? What intrapersonal and interpersonal skills did we practice or develop? and How can we apply those skills in future situations?"

SUMMARY

In this chapter, we explored the dual foci of MP5, choosing efficient tools and using them appropriately, both of which allow students to develop skills in integrity, adaptability, and responsible decision-making, as well as setting goals, taking initiative, communication, and collaboration competencies. We explored two classroom examples of teachers who plan and implement lessons with content and practice objectives, high-quality tasks, collaborative learning strategies, and a balanced approach to assessment.

Questions to Think About

1. How do I model the strategic use of tools, and how do I encourage students to think critically about their choices?

2. In what ways do I create opportunities for students to reflect on their feelings or frustrations when using tools?

3. How do I prompt students to notice and acknowledge their growth in using tools strategically?

4. How did I support students in making decisions about which tool was most appropriate for a given problem?

5. In what ways do I help students understand the importance of perseverance and trying different approaches when a tool doesn't work?

Actions to Take

1. Identify tools within your grade level content standards that students should be exploring, selecting, and using appropriately and strategically (consider concrete manipulatives, pictures, technology, etc.).

2. Collaborate with colleagues to anticipate benefits and limitations of these tools for classroom and life experiences.

3. Explore your curriculum resources and/or adapt your mathematical tasks to encourage the exploration of tools.

4. Select two to three instructional strategies that promote the competencies that naturally align with MP5—integrity, adaptability, and responsible decision-making.

5. Use observation tools, like those provided in this chapter, to monitor students' development of content and practice standards, as well as intrapersonal and interpersonal skills.

CHAPTER 6

FOSTERING SOCIAL-AWARENESS AND SELF-EFFICACY THROUGH ATTENDING TO PRECISION

"ATTEND TO PRECISION" might feel somewhat misleading or vague. Mathematically speaking, precision is often thought of as the exactness of number or the accuracy of the digits in a number. Mathematicians strive to be precise, but Mathematical Practice 6 (MP6) takes a broader perspective on precision to include precise mathematical communication, both orally and in writing. This means the use of clear definitions, appropriate notation, and careful language when discussing mathematical ideas. This includes accurate specification of quantities in numerical answers, graphs, and diagrams. MP6 helps students develop the skills needed to communicate mathematical ideas clearly and accurately, both orally and in writing, fostering a deeper understanding of mathematical concepts (O'Connell & SanGiovanni, 2013).

Mathematics is a language of symbols, and using precise notation is crucial for accurate communication. Even when students compute correctly, they may misuse symbols, like the equal sign, leading to incorrect expressions. For example, when adding 24 + 35, the proper notation involves using separate equalities (e.g., 20 + 30 = 50, 50 + 4 = 54, 54 + 5 = 59) rather than a single incorrect one. Similarly, mathematical vocabulary plays a key role in students' conceptual understanding and ability to communicate effectively. A strong grasp of mathematical vocabulary helps students articulate reasoning, explain strategies, and engage in meaningful discussions, fostering deeper comprehension and supporting academic success (Lee & Herner-Patnode, 2007; Monroe & Orme, 2002; Pierce & Fontaine, 2009).

> **ATTEND TO PRECISION**
>
> *Mathematically proficient students try to communicate precisely to others. They try to use clear definitions in discussion with others and in their own reasoning. They state the meaning of the symbols they choose, including using the equal sign consistently and appropriately. They are careful about specifying units of measure, and labeling axes to clarify the correspondence with quantities in a problem. They calculate accurately and efficiently, express numerical answers with a degree of precision appropriate for the problem context. In elementary grades, students give carefully formulated explanations to each other. By the time they reach high school, they have learned to examine claims and explicitly use definitions.*
> *(CCSSM, 2010)*

There are numerous intrapersonal and interpersonal skills students acquire when engaged in this practice. MP6 calls on students to "use clear definitions in discussion with others and in their own reasoning," which both leverages and supports the development of **communication skills** and **social awareness**. Attending to small, but important details like vocabulary and symbolic notation when communicating ideas enables students to comprehend ideas shared by others and supports students in expressing themselves clearly and convincingly (National Council of Teachers of Mathematics [NCTM], 2000). Furthermore, engaging in MP6 promotes the development of **self-efficacy** and **self-regulation**. Understanding and attending to the precise meaning of vocabulary and symbolic notation helps students develop confidence in their ability to do mathematics and explain it to others.

To effectively integrate these into mathematics lessons, plan for them, explicitly call them out during the lesson, dedicate time to model and discuss them, and reflect on how students applied these actions. Share specific actions in the form of a handout or poster to make visible the integration of social-emotional competencies and MP6 (Table 6.1).

Table 6.1 • *Actions Associated With MP6*

Attend to Precision
• Communicate precisely with others and try to use clear mathematical language when discussing their reasoning • Understand meanings of symbols used in mathematics and label quantities appropriately • Express numerical answers with a degree of precision appropriate for the problem context • Calculate efficiently and accurately

 Download this table at https://companion.corwin.com/courses/wellroundedmathstudent

Engagement in the practices can vary depending on students' age and experience. Regardless of the level, it is crucial to intentionally plan for MP6. Lessons should actively help students connect the practices through the content and relate mathematics to their personal growth.

MERGING CONTENT STANDARDS, MATHEMATICAL PRACTICES, AND SOCIAL-EMOTIONAL COMPETENCIES

When planning and preparing mathematics lessons that integrate MP6 and social-emotional competencies, there are many key decisions to consider. The introductory chapter included a framework for designing mathematics lessons with a social-emotional learning perspective. Rather than adding extra tasks, this framework encourages a thoughtful, intentional approach to planning. It includes guiding questions to support purposeful lesson development. This approach redefines social-emotional competencies not as an "extra" element but as a foundation for deep, meaningful learning in mathematics.

Mathematical Content Standard and Corresponding Mathematics Goal

Start by asking, **"What is the mathematics goal of this lesson?"** Consider this third-grade mathematics standard: "understand that shapes in different categories (e.g., rhombuses, rectangles, and others) may share attributes (e.g., having four sides), and that the shared attributes can define a larger category (e.g., quadrilaterals). Recognize rhombuses, rectangles, and squares as examples of quadrilaterals, and draw examples of quadrilaterals that do not belong to any of these subcategories" (3.G.A.1, CCSSM, 2010). This standard calls on third-grade students to accurately use attributes to classify shapes and recognize that some shapes can be classified into more than one category and a category can be a subset of a larger one. For example, a rectangle is a subset of a parallelogram, which is a subset of a quadrilateral. A mathematics goal aligned to this standard would be for students to understand the relationship among various quadrilaterals by drawing, defining, and labeling them.

At the middle school level, consider the standard that calls on students to "solve multi-step real-life and mathematical problems posed with positive and negative rational numbers in any form (whole numbers, fractions, and decimals), using tools strategically. Apply properties of operations to calculate with numbers in any form; convert between forms as appropriate; and assess the reasonableness of answers using mental computation and estimation strategies" (7.EE.B.3, CCSSM, 2010). This involves students solving contextual problems using rational numbers requiring them to convert between fractions, decimals, and percents as needed. Additionally, students use estimation to justify the reasonableness of their answers. A mathematics goal aligned to this standard would be for students to convert between forms as needed to solve multi-step real-life problems with positive and negative rational numbers in any form.

Mathematical Practice

After unpacking the mathematics in the standard, reexamine its language to identify alignment with the mathematical practices. Ask, **"Which mathematical practice supports engagement in this content standard?"**

Standards that emphasize understanding symbols, terms, and specific attributes (like shapes or fractions) often engage students in MP6. For instance,

- Understand the meaning of the equal sign and determine if equations involving addition and subtraction are true or false. (1.OA.D.7, CCSSM, 2010)
- Solve word problems involving addition and subtraction of fractions referring to the same whole, including cases of unlike denominators,

e.g., by using visual fraction models or equations to represent the problem. (4.NF.B.3d, CCSSM, 2010)

- Use variables to represent numbers and write expressions when solving a real-world or mathematical problem; understand that a variable can represent an unknown number, or depending on the purpose at hand, any number in a specified set. (6.EE.B.6, CCSSM, 2010)

- Create equations in two or more variables to represent relationships between quantities; graph equations on coordinate axes with labels and scales. (HSA.CED.A.2, CCSSM, 2010)

At all levels, students are introduced to new vocabulary and symbols, emphasizing the need for precision. For example, elementary students learn the equal sign as a relationship between quantities, not just an answer indicator (Aisling et al., 2013), laying groundwork for equation-solving in middle school (Knuth et al., 2006).

Students refine their understanding of terms over time, just as in language development. By middle and high school, they apply precise language and symbols in complex mathematical concepts, enhancing their accuracy in explanations, discussions, and representations.

Social-Emotional Competencies

Based on what it means to be engaged in MP6, ask **"What intrapersonal and interpersonal skills are inherent, are needed, and can be further developed for students while engaging in MP6?"** Table 6.1 highlights actions associated with MP6 and outlines actions that enable students to integrate intrapersonal and interpersonal skills with MP6. Embedding these skills into lessons requires an awareness of the connections and intentionality in their incorporation, ensuring they are seen not as separate from the mathematics content, but as integral parts of the learning process. Since students may lack these skills for different reasons, explicitly incorporate, teach, and support their development, even in informal ways.

To achieve this, take advantage of opportunities to emphasize the intrapersonal and interpersonal skills that could naturally correspond to the mathematical task students are asked to complete. For MP6, **self-efficacy** and **self-regulation** are the selected intrapersonal skills, and **communication** and **social awareness** are the targeted interpersonal skills.

Intrapersonal Skills

Chapter 1 explored the connection between **self-efficacy** and perseverance. **Self-efficacy** is a person's belief in their ability to complete a task or achieve a goal (Bandura, 2001). In MP6, self-efficacy is looked at through the lens of building confidence in a student's mathematical abilities. By consistently honing their precision in mathematical skills, students are more inclined to feel self-assured in tackling mathematics problems. This boosted self-efficacy can result in heightened motivation to thrive in mathematics.

In MP6, the focus is on precision, and precision can be or feel tedious, as it takes concentration and great attention to detail. **Self-regulation** is a necessary skill when working with precise tasks. Staying emotionally monitored and behaviorally in control of oneself, especially when the task gets hard, tedious, or requires specificity, is a skill to highlight.

Interpersonal Skills

Utilizing interpersonal skills is an approach to assist students in enhancing **communication skills**. "Attend to precision" in MP6 emphasizes the importance of using clear and accurate mathematical language and representations. While the standard itself does not explicitly mention communication skills, it is closely related to communication. By attending to precision, students learn to articulate their mathematical reasoning clearly and effectively, which is a fundamental aspect of communication in mathematics. MP6 encourages students to express their ideas precisely, which enhances their ability to communicate mathematical concepts to others.

The description of MP6 specifically calls on students to "use **clear** definitions in discussion with others and in their own reasoning." To do this, students need to leverage their **social awareness** skills. It is not uncommon for people to interpret problems and their solutions differently. This means students need to be able to consider the perspectives of other individuals, groups, or the classroom community as a whole and use that understanding to interact with them and **communicate** with them precisely.

Instructional Structures and Engagement Strategies

Shifting toward planning, ask, **"With an eye on our mathematical goal, how will I support social-emotional development as I engage learners in MP6: Attending to precision? What structures, strategies, methods, and/or tools can I use?"** Engaging students in attending to precision involves both explicit instruction and creating opportunities for practice (Elevated Achievement Group, 2020). What follows are specific ways to help students become more adept at using precise mathematical language, accurate representations, and correct notation, along with opportunities for them to practice articulating their reasoning with clarity and accuracy. This can be done by modeling how to provide careful, accurate explanations (both verbal and written) and providing constructive feedback to students as they show and explain their work. To practice attending to precision, plan for peer discussions and engage students in problem-solving activities that require attention to detail. By consistently practicing and reflecting on their written work and communication, students can develop a strong sense of precision in their mathematical thinking and problem-solving skills. Create a classroom environment emphasizing the importance of precise communication in mathematics using various strategies.

Supporting Metacognition

"Attend to precision" is a practice that is metacognitive in nature, making it important to support students in learning to think about their own thinking. Throughout the problem-solving process, there is need for students to attend to precision to efficiently find a clear and accurate solution. However, it is more than finding a correct answer and labeling it correctly. Attending to precision means students use clear and accurate mathematical language to explain what they plan to do, what they are doing, and what they have done. They use accurate symbolic notation to represent the problem situation. They carefully attend to units of measure and labels as they set up and solve a problem. According to the Elevated Achievement Group (2020), to effectively teach, model, and facilitate this skill, teach students to follow these steps:

1. Explain the problem, being careful to use precise mathematics words and symbols.

2. Identify and describe the units of measure in the problem.

3. Outline how you will solve the problem using specific mathematical words and units.

4. While solving the problem, describe each step using exact mathematical language and units.

5. After solving the problem, explain your reasoning both verbally and in writing.

Teaching students to do these things is no small feat. One way to provide a more scaffolded approach is to use Three Writes.

Competency Builder 6.1
Three Writes Routine

Like the Three Reads routine, the Three Writes routine involves students engaging in three passes at solving and responding to a problem, with increasing focus on precision. Students need to attend not only to the details of the problem as they solve it, but also to the precise use of labels, symbolic notation, and vocabulary when explaining their solution.

First, provide students with a problem to solve. During the first write, students put the problem in their own words and consider how to represent it. In the second write, they solve the problem and show their work. During the third write, they revise their work, replacing informal language with more precise words, correcting any errors, and confirming the use of accurate units and labels. This routine helps students build communication and social awareness skills by guiding them to express, improve, and check their understanding, encouraging clear explanations, considering others' viewpoints, building confidence, and self-checking their work.

Using a Graphic Organizer to Focus Attention on Important Aspects of Precision

As you read in Chapter 2, graphic organizers not only prompt students to be more thoughtful and systematic about attending to accurate vocabulary, precise calculations, symbolic notation, units of measure, and labels, but they also act as a tool to help students effectively manage their thoughts, emotions, and behaviors. By breaking down tasks into manageable parts, it is easier to focus, providing clarity and a sense of control. This reduces anxiety and frustration and allows students to feel more confident. The Three Writes routine, combined with the use of graphic organizers, promotes not only mathematical precision, but outlines specific goals and steps for achieving them. Using a graphic organizer in small group settings encourages the sharing of ideas, which in turn supports active listening and respectful discussions.

Competency Builder 6.2
Four Corners Organizer

Use graphic organizers during class discussions (or during small groups to prepare for a discussion) so that students can develop a deeper understanding of the importance of precision in mathematics and learn how to identify their own mistakes and correct them. Graphic organizers like Four Corners help students to recognize the various ways they need to think and rethink the finer details. Provide students with a problem to solve. As students solve the problem, have them respond to the prompts in the organizer (see Figure 6.1). Explain to students that this supports self-regulation and ask them to share how graphic organizers help their learning. Further, discuss how graphic organizers help them to break down and monitor each part of their problem-solving process. By guiding students to define terms, identify units, verify the correct use of symbols, and check calculations, it encourages them to self-check their work at each step, developing habits of careful attention and independent correction.

Figure 6.1 • *Four Corners Organizer*

Clearly define the terms in the problem.	Identify the units being used.
Examine whether the symbols used to solve the problem have been used correctly.	How did you ensure your calculations were accurate?

online resources: Download this figure at https://companion.corwin.com/courses/wellrounded mathstudent

Incorporating Worked Examples

Attending to the details in a worked example is another way students can practice precise communication with one another. They look at the accuracy of the steps as well as the notation. The strategy reduces the load on working memory so students can process information more deeply than they would if they were just using a procedure. Worked examples can be correctly worked, partially worked, and incorrectly worked (Figure 6.2). In correctly worked examples, students review the work, focusing on clarity of explanations and the accuracy of the calculations. They can then apply this to partially worked and incorrectly worked examples. Presenting incorrectly worked examples for students to examine is one way to target misconceptions and construct more accurate conceptual knowledge and procedural skills. However, "[m]erely drawing attention to and correcting errors in calculation does not typically have a large pedagogical payoff for a deep learning of mathematics" (Willingham et al., 2018).

Figure 6.2 • *Partial Sums and Differences Strategy*

PARTIAL SUMS: DECIMALS	PARTIAL SUMS: DECIMALS	PARTIAL SUMS: FRACTIONS
3.7 + 3.8	7.7 + 12.58	$13\frac{4}{5} + 2\frac{1}{10}$
3 + 3 = 6	7 + 12 = 19	13 + 2 = 15
0.7 + 0.8 = 1.5	0.7 + 0.5 = 1.2	$\frac{4}{5} + \frac{1}{10} = \frac{9}{10}$
6 + 1.5 = 7.5	0 + 0.08 = 0.08	$15 + \frac{9}{10} = 15\frac{9}{10}$
	19 + 1.2 + 0.08 = 20.28	

Source: J. J. SanGiovanni et al. (2021, p. 128). Reprinted with permission.

> ## Competency Builder 6.3
> *Worked Examples*
>
> Begin by instructing students on how a particular strategy, method, or procedure works, for example, the strategy of using partial sums or differences when addition or subtracting multidigit numbers (Figure 6.2). This involves a process of decomposing numbers into parts and then adding or subtracting place values from left to right, starting with the largest place value and moving to the smallest (Bay-Williams et al., 2022). Teaching students a strategy like partial sums or differences for adding or subtracting multidigit numbers builds self-efficacy because it gives them a method they can trust and feel confident using. Explain to students that this activity will help them build self-regulation skills. Tell them you will discuss how after the lesson but encourage them to think about and notice how they are using self-regulation skills.
>
> After introducing students to the strategy and exploring how it works, provide them with worked examples (Figure 6.3) to support them in examining the strategy in more detail, recognizing precisely what works, what does not work, and why.

Figure 6.3 • *Worked Examples for Partial Sums and Differences Strategy*

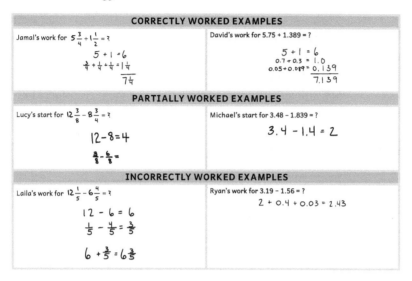

Source: J. J. SanGiovanni et al. (2021, p. 139). Reprinted with permission.

Present students with a worked example. Allow them time to examine the steps. Next, have them discuss each step of the worked example with a peer. Finally, ask them to think about how they used self-regulation skills during this activity. They should recognize that since they needed to follow each step carefully and check their work as they went, they had to slow down and manage their own learning and problem-solving. End by asking them how they use this type of skill outside of mathematics class. When shifting the focus from teaching the "right" way to solve a problem and changing the focus on value of analyzing errors as part of the learning process, students are much more engaged in mathematical practices (Vanoli & Luebeck, 2021). This demonstrates how attending to precision requires looking at the problem more than one time. First, look for an overview and a general approach. Second, look for the details that apply to solving the problem. Third, look back to identify the ways to be more accurate, clear, and concise.

Constructing Claim-Evidence-Reasoning (CER) Data Stories

MP6 states, "In elementary grades, students give carefully formulated explanations to each other. By the time they reach high school, they have learned to examine claims and explicitly use definitions" (CCSSM, 2010). How do they learn to do this? CER data stories provide an opportunity for students to practice making a claim, providing evidence to support the claim, and providing reasoning for why the evidence supports the claim (Tuva Support, 2021).

Competency Builder 6.4
CER Data Stories

Give students a graph and ask them to make a claim based on their interpretation. Ask them to provide evidence from the data in the graph and then explain why it supports the claim. As they formulate a claim, they are careful about attending to specifying units of measure recognizing the labeling axes and use those to clarify the correspondence with quantities in a problem. For example, consider the "Online Platform Usage in the U.S. by Age Group" graph in Figure 6.4.

Figure 6.4 • *Online Platform Usage in the U.S. by Age Group*

[Line graph showing usage percentage (%) on y-axis (0–80) across platforms Facebook, Instagram, Twitter, TikTok on x-axis, with four age group lines: 18–29%, 30–49%, 40–65%, 65+%]

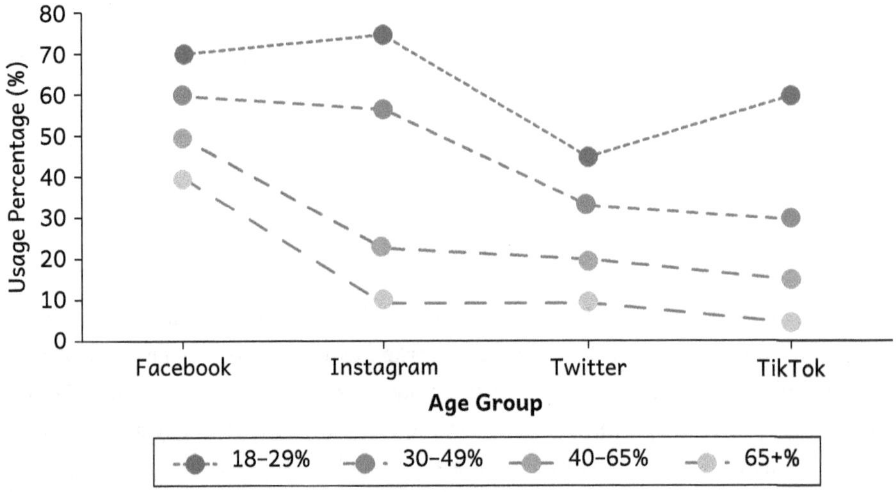

Source: Pew Research Center (2024).

Ask students to make a claim about the graph and explain their reasoning. Explain that having social awareness means they can recognize and consider the perspectives of others, even those that are different from their own. Explain that they develop social awareness by actively listening to others who have different ideas. Then, have students "stand up, pair up" (Kagan & Kagan, 2009) to find a partner who either has a different claim or the same claim with different reasoning. Finally, ask students to write about what they learned from these discussions. Encouraging students to articulate their reasoning and defend their claims cultivates attention to precision. According to Sriraman and Umland (2020), a mathematical argument is "a line of reasoning that intends to show or explain why a mathematical result is true." This skill goes beyond simply formulating an argument; students must also learn to critically evaluate others' reasoning. Given our diverse perspectives, natural disagreements arise, and engaging with multiple viewpoints enhances problem-solving abilities. By fostering these skills, students build social-emotional competencies like social awareness and communication, which align with collaborative problem-solving and thoughtful engagement with differing perspectives.

Some Sources for Graphs That Promote Sense-Making

The New York Times "What's Going on in This Graph" (https://www.nytimes.com/column/whats-going-on-in-this-graph)

Slow Reveal Graphs (https://slowrevealgraphs.com/)

Reflecting on Precision

Encourage students to regularly reflect on their attention to precision by using this prompt: "I can communicate precisely by . . . ." For example, they might write "I have defined all terms clearly, used the appropriate units and labels, and checked my calculations accuracy." Additionally, students could use a checklist (Figure 6.5) that provides scaffolding and breaks down the complexity of a task into manageable parts. We can tell students that this checklist not only guides them through each step of the process, but also supports self-regulation, helping them monitor their progress and ensure that each component of the task is scaffolded in a way to help them monitor themselves.

Figure 6.5 • *Reflecting on Precision*

Looking back at my work . . .
☐ Did I calculate accurately?
☐ Is my answer reasonable?
☐ Did I calculate efficiently? (Is there a way to do this in fewer steps?)
☐ Did I use an appropriate degree of precision when reporting my answer?
☐ Did I use the correct symbolic notation?

(Continued)

(Continued)

When I explained my work . . .
☐ Did I define all terms clearly?
☐ Did I specify units of measure?
☐ Did I clearly label the mathematical model with the correct terms?
☐ Did I explain the meaning of the symbols I used appropriately?

Download this figure at https://companion.corwin.com/courses/wellroundedmathstudent

Ask students how these steps support them in redoing a step, rechecking previous work, or trying a different approach. Ask, "How did this help you identify what went well and where precision could be improved next time?"

Competency Builder 6.5
Prompting Reflection

Tell students that reflecting on their learning helps them to develop self-efficacy, which is their belief in their own ability to accomplish a task or achieve a goal. Explain that reflection helps a person build confidence through their past successes as well as learn from mistakes. When students communicate orally or in writing during mathematics class, you can provide reflection prompts to help them to think critically about when and how precision is essential in various situations. Begin by identifying lessons where you integrate real-world examples where precision is crucial or when students are engaged in reasoning and constructing arguments. Near the end of the lesson, present students with one or more of the following prompts. Have them write down their response and/or discuss their response with a peer. Listen or read individual responses as a means of formatively assessing to what degree and in what ways your students are attending to precision.

- I used precise vocabulary when I . . .
- I improved my use of symbols when I . . .
- One way I made my work more accurate . . .
- I attended to precision by . . .

CHAPTER 6 • FOSTERING SOCIAL-AWARENESS AND SELF-EFFICACY

> Reflecting on how they use precision, students practice self-regulation as they check their own attention to detail and accuracy. The prompts encourage communication by guiding students to explain their thought processes and how they improved their work, helping them learn to share their reasoning clearly. This practice builds self-efficacy as students see their progress in using precise vocabulary, symbols, and accuracy, which boosts their confidence in handling mathematics tasks effectively.

Assessing Interpersonal and Intrapersonal Skills

Last, ask, **"How will I assess students' progress toward the mathematics goal of this lesson, their engagement in the mathematical practice standard, and their ability to use and continue to develop social-emotional competencies? How will I provide feedback? How will I build in an opportunity for students to reflect on the development of intrapersonal and interpersonal skills?"** As described in the introduction, various options are available, but in the following scenarios, the teacher is actively listening and observing students and engaging with them.

To establish assessment criteria, identify actions you want to promote or instill on the part of the student. Table 6.2 illustrates student actions that can be used to create assessment criteria. There are also actions on the part of the teacher that will, without taking over, support and prompt student actions.

Table 6.2 • *Student and Teacher Actions*

Student Actions	Teacher Actions
• Use mathematical terms, both orally and in written form, appropriately. • Use and understanding the meanings of mathematics symbols that are used in tasks. • Calculate accurately and efficiently. • Understand the importance of the unit in quantities.	• Consistently use and model correct content terminology. • Expect students to use precise mathematical vocabulary during mathematical conversations. • Question students to identify symbols, quantities, and units in a clear manner.

Source: Kansas State Department of Education (2017, p. 8).

THE WELL-ROUNDED MATH STUDENT

There are several ways to assess student progress toward developing proficiency with MP6 and social-emotional competencies. When students are engaged in activities like those shared in the competency builders, you can use whole class or individual observation tools (Tables 6.3 and 6.4) to record evidence of what you see and hear. Embedding self-reflection into student awareness supports their understanding of emotions and actions. As they attend to their emotions and actions, students develop the ability to regulate their emotions and actions in alternate settings. To support this development, you may focus on student reflection and assessment of their intrapersonal and interpersonal skills (Table 6.5). Use the following prompts during or after a lesson to guide student reflection. These same prompts can be given to students to help them reflect on their own learning.

- How did you practice **self-efficacy** in today's mathematics activities? Provide at least one example.

- How did you demonstrate **self-regulation** today? Provide at least one example.

- Share two examples of your **communication skills** today. Were these interactions positive or negative? If they were positive, what made them positive? If they were negative, how could you reframe your thinking to make them positive?

- How did you demonstrate **social awareness** in today's lesson? Provide at least one example.

Table 6.3 • *Whole Class Observation Tool*

Name	Mathematics Goal	Engagement in Practice Standard *Attend to precision*	Intrapersonal Competency *Self-efficacy*	Interpersonal Competency *Social awareness*

CHAPTER 6 • FOSTERING SOCIAL-AWARENESS AND SELF-EFFICACY

Name	Mathematics Goal	Engagement in Practice Standard Attend to precision	Intrapersonal Competency Self-efficacy	Interpersonal Competency Social awareness

Note: *Progress will be marked using 0–No evidence, 1–Little evidence, 2–Adequate evidence*

Download this table at https://companion.corwin.com/courses/wellroundedmathstudent

Table 6.4 • *Individual Student Observation Tool*

Name of student:

Mathematics Goal	No Evidence	Little Evidence	Adequate Evidence
Engagement in the Practice	**No Evidence**	**Little Evidence**	**Adequate Evidence**
Precise communication			
Accurate use of mathematical symbols			
Social-Emotional Competencies	**No Evidence**	**Little Evidence**	**Adequate Evidence**
Self-efficacy			
Social awareness			

Note: *Progress will be marked using 0–No evidence, 1–Little evidence, 2–Adequate evidence*

Download this table at https://companion.corwin.com/courses/wellroundedmathstudent

Table 6.5 • *Self-Assessment Checklist*

Social-Emotional Competency	Not Sure	Not Yet	Getting There	Got It
Self-efficacy				
Self-regulation				
Communication				
Social awareness				
Other skills used:				

online resources Download this table at https://companion.corwin.com/courses/wellroundedmathstudent

In the following vignettes, observe both interpersonal and intrapersonal skills in action and the importance of students attending to precision, which is reflected in the teachers' encouragement and the students' problem-solving efforts. The vignettes specifically target the portion of the lesson related to MP6.

LOOKING AT EXEMPLARS IN ACTION

Now that the merging of content standards, mathematical practice standards, and social-emotional competencies has been explored, let's look more closely at an elementary and a secondary example, using the standards and competencies focused on in this chapter.

Mx. Santos and Their Third-Grade Class

Mx. Santos presented their class with a problem focused on the mathematics standard "understand that shapes in different categories (e.g., rhombuses, rectangles, and others) may share attributes (e.g., having four sides), and that the shared attributes can define a larger category (e.g., quadrilaterals). Recognize rhombuses, rectangles, and squares as examples of quadrilaterals, and draw examples of quadrilaterals that do not belong to any of these subcategories" (3.G.A.1, CCSSM, 2010).

Mx. Santos begins the lesson saying, "Class, today we are going to talk about shapes and how we can classify them based on their attributes. First, what is an attribute?"

The class sits silent. Mx. Santos says, "An attribute is a special characteristic or feature that helps us describe or identify something. For example, we are going to look at different shapes and their attributes could include the number of sides or the length of its edges." They continue, "Let's put that word on our Word Wall."

Mx. Santos asks the class, "Can anyone tell me what a quadrilateral is?"

Remy responds, "It's a shape with four sides!"

"And what do we call those sides?" Mx. Santos asks. Francesca says eagerly, "Edges!"

"That's right, Francesca," Mx. Santos says. "A quadrilateral is a shape with four edges. Let's look at some examples of shapes with four edges." They project an image of a rhombus, rectangle, and a square. "What do you notice about these shapes?"

Gabriella says, "They all have four sides, I mean edges."

Chloe adds, "And they all have four angles."

Students can name each of the shapes. Mx. Santos explains that while these shapes share the attribute of having four edges, they also have other unique characteristics that distinguish them from each other.

"Now," says Mx. Santos, "think about other shapes that have four edges but are not rhombuses, rectangles, or squares. Please draw some other quadrilaterals."

The students take out their notebooks and begin drawing shapes like trapezoids and kites. As the students work, Mx. Santos walks around the room, asking students to explain their drawings and how they fit the definition of a quadrilateral. ●

Mx. Santos guided students' understanding of quadrilaterals by using precise language, drawing examples on the board, and engaging students in discussing each shape's attributes. This approach allowed students to visualize and apply the concept practically, building their confidence in identifying and classifying shapes. Mx. Santos encouraged creative thinking through open-ended questions, helping students use precise mathematical language and develop communication skills as they explained their ideas.

In fostering an environment of curiosity and self-efficacy, Mx. Santos empowered students to explore different shapes and their characteristics. By setting clear goals and emphasizing new vocabulary, they supported students' self-regulation and focus, allowing them to stay engaged and build a strong understanding of quadrilateral attributes.

Mrs. Sparkman and Her Seventh-Grade Class

Mrs. Sparkman is teaching the mathematics standard "solve multi-step real-life and mathematical problems posed with positive and negative rational numbers in any form (whole numbers, fractions, and decimals), using tools strategically. Apply properties of operations to calculate with numbers in any form; convert between forms as appropriate; and assess the reasonableness of answers using mental computation and estimation strategies" (7.EE.B.3, CCSSM, 2010). Mrs. Sparkman poses a challenging mathematics problem that involves multiple steps and rational numbers. She wants to engage her students in a real-life scenario to make the problem more meaningful. She projects this task from the Illustrative Mathematics (n.d.) website on the board:

> *Anna enjoys dinner at a restaurant in Washington, D.C., where the sales tax on meals is 10%. She leaves a 15% tip on the price of her meal before the sales tax is added, and the tax is calculated on the pre-tip amount. She spends a total of $27.50 for dinner. What is the cost of her dinner without tax or tip?*

Mrs. Sparkman begins, "Please take a moment to read the problem carefully. What do we need to find?"

Remy is the first to answer. "We need to find out how much Anna's dinner costs without the tax or tip."

"Exactly," Mrs. Sparkman responds. "How can we approach this problem? Break it down step-by-step. Who can tell me where to start?"

"We need to find the cost of meal without the tip," says Rohan. "We can work backward," says Neilander.

"How are we going to do that?" asks Mrs. Sparkman.

"Since the tip is 15%, and it is figured in before tax, we can find the cost before the tip by dividing the total cost by 1 plus 15%," says Rohan.

"Excellent!" says Mrs. Sparkman. "What equation can we write, if x is the cost before the tip?"

Chloe says, "$x + 0.10x + 0.15x = 27.50$."

"How did you get that?" asks Remy. Chloe replies, "If x is the cost before the tip, then $0.10x$ is the sales tax, and $0.15x$ is the tip."

"Good job, Chloe," says Mrs. Sparkman. "You have explained with precision what each term represents. Now let's solve this equation together."

Mrs. Sparkman then tasks students with solving the equation, following the order of operations. When it was clear most students arrived at a solution, Mrs. Sparkman asks the students for the cost of Anna's dinner before tax or tip, which is $22.

Mrs. Sparkman supported her students' understanding of multi-step real-life problems with decimals and applied properties of operations. She clearly helped students focus on their thinking when she asked them how to approach the problem and break it down into smaller, manageable parts. Mrs. Sparkman modeled how to solve the equation and provided students examples as she reinforced the concept of percentages, tax, and tips.

Mrs. Sparkman supported students' intrapersonal and interpersonal growth within MP6 by nurturing their self-efficacy. She created an empowering environment where students felt capable and supported. Curiosity was supported as Mrs. Sparkman used a real-life problem as they explored a concept in more depth. As Mrs. Sparkman encouraged her students to determine what they need to solve the problem, she supported their curiosity as they made sense of the problem. Additionally, when Neilander suggested working backward to find the cost, Mrs. Sparkman encouraged all students to think creatively.

Mrs. Sparkman encouraged students to use communication skills when she emphasized clear and accurate expressions and attention to detail, and students attended to self-regulation as they were precise and careful in their work. Chloe explained her equation with precision and broke down each component showing a clear understanding and her ability to articulate mathematical concepts effectively. ●

Reflection

To further enrich and reinforce the lesson, incorporate reflection to deepen learning. Reflective practice is crucial for the effectiveness of any lesson, as it aids in the growth and enhancement of both the teacher and the learner. It involves both you and students evaluating the lesson's effectiveness and identifying areas for future improvement. This practice naturally fosters self-awareness and the synthesis of learning, engaging in higher order thinking skills.

When reviewing the lesson, it's important to assess students' grasp of their mathematics competencies and understanding of intrapersonal and interpersonal skills. Engaging in a class discussion can be highly beneficial. There's always room for improvement, with a focus on identifying and discussing positive aspects while also considering ways to enhance the learning experience.

SUMMARY

This chapter examined how teachers leverage MP6 and draw on the social-emotional competencies to enhance their classroom lessons. MP6 emphasizes precise communication, including using clear language, notation, and representations to convey mathematical thinking, fostering a deeper understanding and effective communication. All of these are crucial for students to articulate their reasoning and engage in productive discourse, which will ultimately enhance their mathematical understanding and communication skills. Aligning mathematics lessons with MP6 by using standards like 5.NF.B.5.b and 7.EE.B.3 will ensure that students understand concepts including multiplying fractions and solving multi-step problems with rational numbers, while also emphasizing the significance of precise mathematical communication. The integration of intrapersonal and interpersonal skills, such as **self-regulation, self-efficacy, curiosity,** and **communication** all focus on the importance of clear and accurate communication and the development of confidence and curiosity in mathematics. Engaging students in clear and accurate mathematical communication, promoting reasoning and problem-solving skills through high-quality tasks, and integrating precision with other mathematical practices will develop students' skills in attending to precision.

In the classroom narratives, Mx. Santos's fifth-grade class engaged in a mathematics lesson on recognizing and categorizing shapes based on shared attributes, focusing on self-regulation and self-efficacy. In Mrs. Sparkman's seventh-grade class, students solved multi-step real-world mathematical problems involving positive and negative rational numbers, demonstrating precision and the real-world application of mathematics concepts.

Questions to Think About

1. As a teacher of mathematics, how can you integrate social-emotional competencies into your lesson planning process and in your classroom routines? Provide some examples of how this might look in practice.

2. Think of a specific mathematics lesson that would incorporate MP6. Which intrapersonal or interpersonal skills would enhance the lesson?

3. How do you ensure active integration of MP6 and social-emotional competencies into your mathematics lessons?

4. Reflect on a recent mathematics lesson you taught. How could you have integrated self-efficacy, curiosity, communication skills, and self-regulation in the preparation and planning to positively impact your teaching and student learning?

5. What are other intrapersonal and interpersonal skills you could incorporate in this mathematical practice? What are other ways to naturally amplify social-emotional competencies within the lesson?

Actions to Take

1. Be intentional about reflecting with your students about the skills they developed throughout the lesson. Pose the following questions to guide the discussion:
   - What mathematics skills did we develop today?
   - What other skills did you/we use to practice or learn this concept?
   - What is the value of learning this concept individually/together?
   - What intrapersonal or interpersonal skills did you apply in this lesson to strengthen students' understanding?

2. Reflect on how you implemented MP6 in the classroom.
   - What strategies did they use?
   - What challenges did they face?
   - How did they persevere through those challenges?

3. Discuss with your colleagues how you merge content, practices, and social-emotional competencies in your mathematics classroom, specifically related to MP6.

CHAPTER 7

LEVERAGING ADAPTABILITY TO FIND AND USE STRUCTURE

THE TERM *STRUCTURE* COMES FROM THE LATIN WORD *STRUCTURA*, which means "a fitting together" or "a building" (Vocabulary.com, n.d.). While people commonly associate structures with things like houses, sentences, or skeletons, they are also fundamental in mathematics. In this field, structure pertains to the organization and relationships among numbers, quantities, and patterns, as well as the logical frameworks used. Structure is integral to various areas of mathematics, including the number system, geometry, algebra, and logical reasoning.

> **LOOK FOR AND MAKE USE OF STRUCTURE**
>
> *Mathematically proficient students look closely to discern a pattern or structure. Young students, for example, might notice that three and seven more is the same amount as seven and three more, or they may sort a collection of shapes according to how many sides the shapes have. Later, students will see 7×8 equals the well-remembered $7 \times 5 + 7 \times 3$, in preparation for learning about the distributive property. In the expression $x^2 + 9x + 14$, older students can see the 14 as 2×7 and the 9 as $2 + 7$. They recognize the significance of an existing line in a geometric figure and can use the strategy of drawing an auxiliary line for solving problems. They also can step back for an overview and shift perspective. They can see complicated things, such as some algebraic expressions, as single objects or as being composed of several objects. For example, they can see $5 - 3(x - y)^2$ as 5 minus a positive number times a square and use that to realize that its value cannot be more than 5 for any real numbers x and y. (CCSSM, 2010)*

THE WELL-ROUNDED MATH STUDENT

This type of work requires intentionality. Specifically, when students "look for and make use of structure" in mathematics, they engage in specific actions (Table 7.1).

Table 7.1 • *Actions Associated With MP7*

Action	Examples
Use what is known to solve unfamiliar problems	Students see that adding fractions with unlike denominators that are natural numbers (e.g., $\frac{1}{2} + \frac{1}{3} = \frac{3}{6} + \frac{2}{6} = \frac{5}{6}$) can help them make sense of adding fractions with variables in the denominator (e.g., $\frac{1}{4x} + \frac{3}{y} = \frac{y}{4xy} + \frac{12x}{4xy} = \frac{12x+y}{4xy}$).
Break apart complicated structures into easier and more accessible components	Students see when adding 8 + 5 they can break 5 into 2 and 3 to make a 10 (i.e., 8 + 2 = 10, 10 + 3 = 15). When solving $x^2 + 8x + 15 = 0$, students see 15 as 5 × 3 and see 8 as 5 + 3 to solve by factoring.
Step back and embrace another perspective by identifying relationships within mathematical expressions/equations and rewriting them	Students see that 7 + 6 + 3 can be written as 7 + 3 + 6 and then 10 + 6. Students see that $2x + 3y - 4x - 8y + 7x$ can be written as $5x - 5y$ and $5(x - y)$, or that $2y = 3x + 10$ can be rewritten as $y = \frac{3}{2}x + 5$.
Focus on the structure within mathematics and use this to connect concepts and representations (objects, tables, graphs, visuals, expressions, etc.)	Students use link cubes to see 4 + 5 = 9 and can then see that 9 − 4 = 5 and 9 − 5 = 4. Students use base-ten blocks to see the value of the digits in a decimal number (e.g., 3.24). =1 = 0.1 = 0.01 3.24 =

CHAPTER 7 • LEVERAGING ADAPTABILITY TO FIND AND USE STRUCTURE

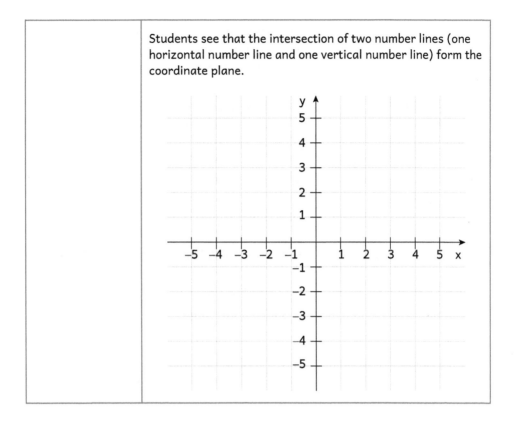

Students see that the intersection of two number lines (one horizontal number line and one vertical number line) form the coordinate plane.

In addition, Mathematical Practice 7 (MP7) has important implications when it comes to procedural fluency. Consider the following content standards:

- Fluently add and subtract within 20 using mental strategies. (2.OA.2, CCSSM, 2010)
- Fluently add and subtract multi-digit whole numbers using the standard algorithm. (4.NBT.4, CCSSM, 2010)
- Compute fluently with multi-digit numbers and find common factors and multiples. (6.NS.B, CCSSM, 2010)
- Solve word problems leading to the equations of the form $px+q=r$ and $p(x+q) = r$, where p, q, and r are specific rational numbers. Solve equations of these forms fluently. (7.EE.B.4a, CCSSM, 2010)
- Perform arithmetic operations on polynomials. (HS.APR.A, CCSSM, 2010)

Notice how the term *fluently* is used in these standards. Procedural fluency means being able to use procedures efficiently, flexibly, and accurately (National Council of Teachers of Mathematics [NCTM], 2014; National

Research Council [NRC], 2001). Solving problems efficiently involves choosing the right strategy. This means, "Of all the available strategies, the one the student opts to use gets to a solution in about as many steps and/or about as much time as other appropriate strategies" (Bay-Williams & SanGiovanni, 2021, p. 4). This requires students to first examine the structures within the problem they are working with to determine the best approach. For example, consider 199 + 198. Stepping back, a student might see that 199 is 1 away from 200 and 198 is 2 away from 200, and if 200 + 200 = 400, then 199 + 198 would equal 3 less than that (397). Or they might recognize that 199 + 1 = 200 and borrow 1 from 198, resulting in 197. Then 200 + 197 = 397. Applying procedures flexibly means students can trade out a strategy that is not working or adapt a strategy to fit the numbers in the problem, as well as apply a strategy to a new problem type (Bay-Williams & SanGiovanni, 2021, p. 4). This requires that students shift from a narrow perspective on the use of procedures in an isolated way to one based on the structure of the content. According to the National Research Council (2001),

> Students need to see that procedures can be developed that will solve entire classes of problems, not just individual problems. By studying algorithms as "general procedures," students can gain insight into the fact that mathematics is well structured (highly organized, filled with patterns, predictable) and that a carefully developed procedure can be a powerful tool for completing routine tasks. (p. 121)

When students select and apply an appropriate strategy based on the structure of a mathematical problem and adapt their approach as necessary, they combine MP7 with social-emotional competencies.

MP7 encourages students to recognize patterns and structures within mathematical problems, helping them to become flexible thinkers. By identifying underlying structures, students demonstrate **adaptability** as they solve problems in new and varied contexts. This enhances their ability to tackle unfamiliar or new situations with confidence. MP7 also requires **self-regulation**, prompting students to plan, monitor, and adjust their approach when solving problems. As students focus, break down complex problems into manageable pieces, and reflect on their strategies and solutions, they further develop their ability to self-regulate. Overall, involving students in MP7 helps them build and strengthen their ability to adapt and be flexible while identifying mathematical patterns and structures to solve problems.

MERGING CONTENT STANDARDS, MATHEMATICAL PRACTICES, AND SOCIAL-EMOTIONAL COMPETENCIES

When selecting mathematical standards and goals that integrate MP7 into classroom lessons, consider the natural connections between MP7 and intrapersonal and interpersonal skills. In the introduction and subsequent chapters, you have learned a framework for designing lessons that emphasize social-emotional learning. As we have highlighted, this framework should not add to the workload but should promote a thoughtful approach to lesson planning. The following guiding questions will help you integrate social-emotional skills into your lessons as an essential part of meaningful mathematics learning rather than as an additional task.

Mathematical Content Standard and Corresponding Mathematics Goal

Begin by asking, **"What is the mathematics goal of this lesson?"** Consider the fifth-grade standard in which students are asked to "[e]xplain patterns in the number of zeros of the product when multiplying a number by powers of 10 and explain patterns in the placement of the decimal point when a decimal is multiplied or divided by a power of 10. Use whole number exponents to denote powers of 10" (5.NBT.A.2, CCSSM, 2010). The concept of multiplying by 10 is often approached as simply "moving the decimal." While this trick may feel like it helps students, it does little to connect with students' structural understanding of the base-ten place value system. Rather than a focus on "moving the decimal," instruction can focus conceptually on "shifting the digits." One mathematics goal for a lesson addressing this standard would be for students to recognize and use the structure of the base-ten number system when multiplying and dividing by powers of 10. Specifically, students will recognize the value of a digit increases 10 times when moving left on the place value chart and decreases by a factor of $\frac{1}{10}$ when moving to the right. For example, the digit 2 in the hundreds place (200) would be 100 (10 × 10) times greater than a 2 in the ones place. Students will then be able to explain how this applies to multiplying or dividing by a power of 10.

At the high school level, consider the standard that asks students to "graph functions expressed symbolically and show key features of the graph, by hand in simple cases and using technology for more complicated cases"

(HSF-IF.C.7, CCSSM, 2010). Students should be able to describe and sketch key features of different functions represented graphically and algebraically. They predict behavior by focusing on specific characteristics within graphs or functions or stepping back to view the entire graph or function as necessary. These actions happen iteratively, bouncing between overviews and narrow foci to determine and predict function shapes and equations. Therefore, one mathematics goal would be for students to use structure of functions to determine graphs and algebraic equations.

Mathematical Practice

Next ask, **"Which mathematical practice supports engagement in this content standard?"** In the fifth-grade standard, notice the language "explain patterns in the number of zeros" and in the high school standard "show key features of the graph." Phrases like these call on students to examine patterns and representations to delve into the core and uncover underlying structures. Notice these standards also call on students to take a more generalized view. They are to consider the whole of something but also decompose it into simper parts. This type of work aligns well with MP7. Within mathematics standards that align with MP7, words such as *identify*, *analyze*, *compare*, *model*, *describe*, *apply*, *understand*, and *relate* are used. Other examples of these standards include

- Understand that the three digits of a three-digit number represent the number of hundreds, tens, and ones. (2.NBT.A.1, CCSSM, 2010)
- Use similar triangles to explain why the slope *m* is the same between any two distinct points on a non-vertical line in the coordinate plane. (8.EE.A.6, CCSSM, 2010)

Notice that connections to the structure and concept are often embedded within content standards. For example, the second-grade standard is based on the structure of the base-ten number system, drawing on place value and highlighting language representing values such as ones, tens, and hundreds. The eighth-grade standard connects similar right triangles and nonvertical lines in the coordinate plane to slope. Students use the graph of a line to construct triangles between two points and compare the sides, recognizing there is a constant rate of proportion between them, to understand the slope is the same between any two points. In general, these standards call on students to leverage existing structures to help them solve problems and to build more sophisticated concepts.

Social-Emotional Competencies

Based on what it means to be engaged in MP7, ask, **"What intrapersonal and interpersonal skills are inherent, are needed, and can be further developed while students engage in MP7?"**

The integration of skills requires intentional connections and purposeful implementation, ensuring that social-emotional competencies are not seen as separate from the lesson content, but rather as inherent aspects of learning. Therefore, you should explicitly teach, draw from, and support these skills, even informally, to ensure their development. To achieve this, key components of the lesson will emphasize the social-emotional competencies that naturally correspond to the mathematical tasks students are asked to complete. Prioritizing intrapersonal skills assists students in enhancing self-awareness and self-development in different situations. For this mathematical practice, **self-regulation** is the targeted intrapersonal skill, and **adaptability** is the targeted interpersonal skill.

Intrapersonal Skills

Catching students in the moment and naming their positive social-emotional competencies requires no additional effort but has great payoff. As students demonstrate **self-regulation** while attending to mathematical structure, teachers can highlight their actions and make connections explicit. For instance, a student who was once impulsive and struggled with self-control can learn to think more deeply, monitor, and stay in control of their emotions and responses. Acknowledge a student in the moment and explicitly identify their correct behavior, such as by saying, "I like how you put your pencil down to avoid erasing your work as you examined the structure of this complex problem." This supports and affirms the student's self-control and efforts, making it easier for them to develop positive feelings toward self-regulation and replicate this behavior in the future.

Interpersonal Skills

Adaptability, or the ability to adjust and grow with new information, can be amplified within MP7. As students attempt and see answers, they can compare responses and adjust their problem-solving and understanding. This skill can be implemented in the classroom and encouraged in other areas of life. The classroom is an excellent environment to foster adaptability and encourage students to apply it in their daily lives. Incorporate prompts like "How have you adjusted your thinking as you learned more information?" or "Describe a time when you adapted your approach and viewed the situation differently."

Instructional Structures and Engagement Strategies

Shifting toward planning, ask, **"With an eye on our mathematics goals, how will I support social-emotional development as I engage learners in MP7: Look for and make use of structure? What structures, strategies, methods, and/or tools can I use?"**

When planning lessons to develop MP7, adaptability and self-regulation rely on strategies, routines, and tools that help students identify underlying structures and discern patterns. The goal is to enable students to understand the overall concept by breaking down complex tasks into manageable steps and using their existing knowledge to tackle new challenges. For example, students proficient in the compensation strategy for whole numbers can apply it to fractions. MP7 can be challenging because students may struggle to identify patterns or structures. This process develops self-regulation as they learn to manage their behavior, emotions, and thoughts while uncovering structures. The following strategies and routines support these mathematics and social-emotional goals.

Finding Structure

Mathematical routines provide a predictable structure that helps students develop a sense of security. It is important to discuss with students the social-emotional benefits of routines, so they recognize and participate meaningfully in them. Repeating routines helps students form positive habits that become more automatic over time. This supports students in managing behaviors and emotions more effectively, reducing stress and impulsivity. The following routines were selected because they also offer opportunities for students to find and make use of mathematical structures while practicing self-regulation.

Competency Builder 7.1
What Do You Notice?

The What Do You Notice? routine enhances social-emotional competencies while supporting MP7 by encouraging students to observe different aspects of a problem and identify various entry points and strategies for solving it. This routine also helps students manage their emotions and behaviors, listen respectfully, consider others' perspectives, and collaborate effectively during peer interactions and discussions.

Sometimes students need a brief lesson to activate their thinking. Teachers can highlight patterns in mathematics by displaying parts of a concept and asking, "What do you notice?" This routine shifts the focus from teachers simply providing information students may have forgotten to actively engaging them in noticing and reasoning. Instead, teachers can leverage MP7, encouraging students to observe structure and patterns in mathematics.

For example, instead of explaining how powers of 10 and exponents are related, the teacher can display three columns on a whiteboard (Table 7.2), initially hiding columns two and three. The teacher points to the first column, asks, "What do you notice about these numbers?" and provides quiet think time. Afterward, students discuss their observations with a neighbor. Following this, the teacher reveals the second column, repeats the question, and allows for more think time and discussion. Finally, the teacher reveals the third column, inviting further noticing and discussion.

Table 7.2 • *Patterns of Ten*

10	10	10^1
100	10 × 10	10^2
1,000	10 × 10 × 10	10^3
10,000	10 × 10 × 10 × 10	10^4
100,000	10 × 10 × 10 × 10 × 10	10^5

Through strategic conversation and student noticing, students recognize patterns in mathematics and connect powers of 10 and exponents. They notice that the exponent corresponds to both the number of zeros in the numeral and the number of times 10 is multiplied. As they gain more information, students adapt their thinking and practice self-regulation by maintaining effort and staying calm, even when the mathematics becomes challenging or tedious.

Competency Builder 7.2
Card Sort

The Card Sort routine enhances social-emotional competencies and supports MP7 in two main ways. First, it encourages students to categorize and organize cards from different perspectives, promoting problem-solving from multiple angles and adaptability in grouping information based on new insights. Additionally, it requires students to monitor their progress, verify the logic of their groupings, and reflect on their reasoning.

Card Sorts engage students in understanding the structure of mathematics and foster adaptability by allowing them to focus on specific details or take an overview perspective. At the elementary level, students match cards with related number sentences, number lines, number bonds, and tape diagrams (Figure 7.1). This highlights connections between different forms, promoting understanding and adaptability. Alternatively, they can analyze cards that express numbers in various forms, such as standard, written, unit, numeral, and expanded forms (Figure 7.2), to focus on place value and structure.

Figure 7.1 • *Early Elementary Card Sort Sample Set*

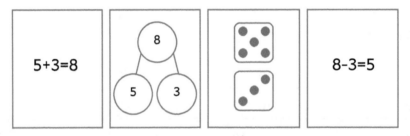

Figure 7.2 • *Upper Elementary Card Sort Sample Set*

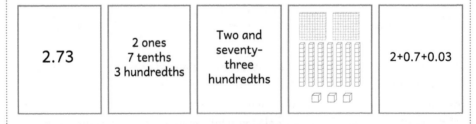

At the secondary level, students match cards with linear equations in different forms, tables, and graphs, to identify parallel or perpendicular lines by examining

CHAPTER 7 • LEVERAGING ADAPTABILITY TO FIND AND USE STRUCTURE

slopes (Figure 7.3). For example, they may recognize an equation's slope and then find the corresponding slope in a graph, table, or other linear models by stating, "This equation has a slope of $\frac{2}{5}$. I know this because the coefficient of x is $\frac{2}{5}$. I need to find another model with a slope of $\frac{2}{5}$. How do I see that in a graph, table, or other linear models?" This activity enhances their understanding beyond equivalence to connections within different forms.

Figure 7.3 • *Secondary Card Sort Sample Set*

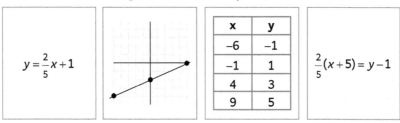

In all levels, students work with partners to match corresponding cards, helping them understand the connections in mathematical concepts. They may need to focus on specific aspects or step back to see the overall picture, sometimes looking at an entire graph, table, number line, number bond, or tape model. This process helps them adapt their view and solve problems while developing self-regulation by managing frustration and effort to continue finding answers. A graphic organizer can help students record their thinking (Table 7.3).

Table 7.3 • *Card Sort Graphic Organizer*

	How do you know these cards are [parallel or perpendicular]? Be specific in your reasoning.
Card ___ matches Card ___	
Card ___ matches Card ___	
Card ___ matches Card ___	
...	

(Continued)

(Continued)

Some Sources for Card Sorts

Card Sorts (https://mathequalslove.net/category/practice-structures/card-sorts/f)

Illustrative Mathematics: Grades 6–8 Instructional Routines (https://hub.illustrativemathematics.org/s/68/68-routines-list)

Illustrative Mathematics: Grades 9–12 Instructional Routines (https://hub.illustrativemathematics.org/s/912/912-routines-list)

Competency Builder 7.3
Cover and Uncover

The Cover and Uncover routine develops social-emotional competencies and supports MP7 by requiring students to adapt their thinking to new aspects of the problem and maintain attention as they work through various elements. The routine asks students to examine the given structure, predict the "covered" part, and then compare their prediction with the uncovered part. This approach promotes strategic thinking by focusing on understanding and number sense, rather than just following procedures.

This strategy builds relational reasoning as a strategy for solving equations. Relational reasoning involves looking at an equation holistically while analyzing the structure of it to identify relationships that lead to finding the value of the variable (Bay-Williams et al., 2022). This routine leverages the structure of a number sentence, rule, expression, or equation by asking students to examine the given structure, predict the quantity for the "covered" part, and continue to cover and uncover until the solution is found.

Cover and Uncover promotes strategic thinking as students can derive rules that leverage number sense and understanding rather than relying solely on mimicked, step-by-step procedures. Figure 7.4 illustrates how this routine works. Adapt the equations to be developmentally appropriate for your students.

CHAPTER 7 • LEVERAGING ADAPTABILITY TO FIND AND USE STRUCTURE

Figure 7.4 • *Cover and Uncover Mathematics Equation*

$4(2x - 3) + 2 = 30$			
✋ $+ 2 = 30$ $4(2x - 3) = 28$	4 ✋ $= 28$ $2x - 3 = 7$	✋ $- 3 = 7$ $2x = 10$	2 ✋ $= 10$ $x = 5$

Source: *(hand icon):* Istock.com/bortonia

Display the equation on the board, covering part of it so students see that some quantity plus 2 equals 30. They conclude the "covered" part must equal 28. Uncover it to reveal $4(2x - 3) = 28$. Write this on the board. Next, cover $2x - 3$ and ask, "4 times what equals 28?" Students conclude it must equal 7. Uncover to show $2x - 3 = 7$. Repeat by covering $2x$ and ask, "What number minus 3 equals 7?" They find it's 10. Finally, cover x and ask, "2 times what equals 10?" They find X equals 5. Practice this process with the class or in pairs.

As students engage, emphasize adaptability by naming and defining it. Explain that being adaptable means adjusting thinking and behavior based on new information. This routine also teaches how to break down a task into manageable steps, an essential skill for self-regulation. Discuss how this strategy can be applied beyond mathematics, like breaking down overwhelming tasks into smaller parts, like creating a daily schedule.

Adapting Thinking

As stated previously, mathematical routines provide consistency while students develop social-emotional competencies. These structures allow them to focus on learning and adapting strategies without feeling overwhelmed. They give students a place to start and a way to go. Routines can introduce challenges that gradually increase in complexity, helping students build confidence and adapt strategies as they progress. Additionally, these mathematical routines involve collaboration, allowing students to share strategies and ideas, exposing them to diverse perspectives and approaches. The following routines support students adapting their thinking as they attend to and make use of structure.

Competency Builder 7.4
Which One Doesn't Belong?

The Which One Doesn't Belong? routine develops social-emotional competencies and supports MP7 by highlighting differences within related mathematical concepts and offering opportunities to understand others' perspectives. This routine allows us to draw attention to the structure within mathematical objects such as graphs, tables, equations, representations, and terms, and as a result students are more likely to attend to the specific attributes of those objects. To begin, display three to four related images or mathematical objects, ensuring at least one reason why an object does not belong, though there may be multiple reasons (Figure 7.5).

Figure 7.5 • *"Shapes" Which One Doesn't Belong?*

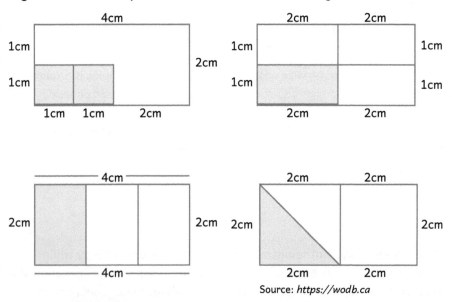

Source: *https://wodb.ca*

Begin by asking, "Which one doesn't belong, and why?" Give students independent think time, then have them discuss with a partner before sharing with the class. Ask, "Did anyone else see it that way?" and "Who saw something different?" to encourage diverse perspectives.

This routine fosters adaptability as students adjust their thinking based on new insights from others. You can make this process explicit by explaining adaptability as the ability to change thinking based on new information, using questions like the following:

- How did your understanding grow or adapt with this new information?
- After hearing other perspectives, how did your understanding grow and adapt?

This activity is ideal for introducing new concepts or summarizing learning. Conduct it while standing around the images to actively engage students and promote thinking.

Competency Builder 7.5
Math Talks

The Math Talks routine develops social-emotional competencies and supports MP7 by exposing students to multiple problem-solving strategies, encouraging openness to different approaches and adaptability (Joswick & Taylor, 2022). It requires careful listening, engagement, and focus, aiding in maintaining attention over time. Additionally, managing frustration and staying calm during challenging problems supports self-regulation. Students need to think flexibly and mentally about numbers, operations, and procedures. Number Talks (Humphreys & Parker, 2015), or Math Talks, promote this by focusing on various ways to mentally find a solution. Teachers can highlight the structure of expressions, number decomposition, and others' perspectives.

To engage students in this routine, display each equation or expression one at a time, provide individual think time, encourage student discourse with a small group or partner, and then engage in a whole class discussion where students share their strategies and problem-solving methods. Teachers enhance the routine's effectiveness by encouraging students to understand and "borrow" another student's strategy.

Here are two examples that support MP7 as students make use of structure (Table 7.4).

Table 7.4 • *Math Talk Examples*

Elementary Example	Secondary Example
4×2	$x + 1 = 5$
4×10	$2(x + 1) = 10$
4×20	$3(x + 1) = 15$
4×18	$10(x + 1) = 50$
	$500 = 100(x + 1)$

(Continued)

(Continued)

During this routine, teachers can accentuate diverse solution paths and spotlight specific strategies, using questions such as

- Do you agree or disagree? Why?
- Did anyone have the same strategy but would explain it differently?
- Who can restate [student's] reasoning in a different way?
- Does anyone want to add on to [student's] strategy?
- Did anyone solve the problem in a different way?
- A student from another class said, "_____." What do they mean? Discuss this strategy and its usefulness with someone near you.
- Did you borrow a strategy from someone else? How did that different perspective help?

Sometimes, teachers unintentionally undermine the opportunities for students to attend to patterns and structure within mathematics. For example, teachers may say, "Just apply the multiplication algorithm on paper or with a whiteboard" or "Always distribute when you see a problem like this." This undermines a student's opportunity to self-select strategies and

- Adapt: Can you see this from another perspective?
- Self-regulate: Before you jump in and automatically (multiply/distribute) and solve, can you step back, look at the equation, and notice the structure to solve the problem mentally?

Teachers can further enhance the intrapersonal and interpersonal connections by affirming student actions.

- You solved the problem using another student's idea or perspective. That shows adaptability in your thinking!
- You paused to think about how you might solve this problem without writing it out. Way to demonstrate self-regulation and manage your impulsivity!

Assessing Interpersonal and Intrapersonal Skills

After selecting how students will engage with the content and the competencies to develop, teachers naturally wonder how to assess if learning goals were met. Questions might include **"How will I assess students' progress toward the mathematics goal of this lesson, their engagement in the mathematical practice standard, and their ability to use and continue to develop intrapersonal and interpersonal skills? How will I provide feedback?"** These questions can be answered using a balance of formative and summative assessment strategies. Embedding a variety of observations and self-reflection tools (Table 7.5, Table 7.6, and Table 7.7) can improve emotion management, self-regulation, and adaptability across settings.

Table 7.5 • *Whole Class Observation Tool*

Name	Mathematics Goal	Engagement in Practice Standard *Identifies a pattern or structure*	Engagement in Practice Standard *Makes use of structure to solve problem*	Intrapersonal Competency *Self-regulation*	Interpersonal Competency *Adaptability*

Note: *Progress will be marked using 0–No evidence, 1–Little evidence, 2–Adequate evidence*

Download this table at https://companion.corwin.com/courses/wellroundedmathstudent

Table 7.6 • *Individual Student Observation Tool*

Name of student:

Mathematics Goal	No Evidence	Little Evidence	Adequate Evidence
Engagement in the Practice	**No Evidence**	**Little Evidence**	**Adequate Evidence**
Identifies a pattern/structure			
Makes use of structure to solve problem			
Social-Emotional Competencies	**No Evidence**	**Little Evidence**	**Adequate Evidence**
Self-regulation			
Adaptability			

online resources Download this table at https://companion.corwin.com/courses/wellroundedmathstudent

Observation tools effectively monitor skill development, while self-reflection prompts help students assess their self-efficacy in social-emotional competencies. Teachers can focus on student reflection of intrapersonal and interpersonal skills using these tools.

Observe and listen to students as they work. Ask them questions, such as the following:

- How did you adapt your thinking to another perspective? Provide an example.
- How did you step back for an overview in today's mathematics lesson? Provide an example.
- How did you self-regulate today? If not, when could you have self-regulated? What can you do differently next time?

Record your observations based on students' responses.

Meaningful student reflection supports the development of social-emotional competencies. When students connect their actions to emotions, it supports their social-emotional growth. Use a simple self-assessment checklist to promote that growth.

Table 7.7 • *Student Self-Assessment Quick Check*

Social-Emotional Competency	Not Sure	Not Yet	Getting There	Got It!
Self-regulation				
Adaptability				
Other skills used:				

online resources Download this table at https://companion.corwin.com/courses/wellroundedmathstudent

Assessing students' social-emotional development may seem challenging, but it doesn't require complex tools. Simple, intentional prompts and thoughtful assessments can effectively support this growth.

LOOKING AT EXEMPLARS IN ACTION

Now that we have explored the merging of content standards, mathematical practice standards, and social-emotional competencies, let's look more closely at an elementary and a secondary example, using the standards and competencies focused on in this chapter.

Mrs. Reynolds and Her Fifth-Grade Class

Mrs. Reynolds plans to use a Notice and Wonder routine in her lesson that focuses on having students "[e]xplain patterns in the number of zeros of the product when multiplying a number by powers of 10 and explain patterns in the placement of the decimal point when a decimal is multiplied or divided by a power of 10. Use whole number exponents to denote powers of 10" (5.NBT.A.2, CCSSM, 2010). This will support students in looking for the underlying structure in the pattern of zeros when multiplying and dividing by 10. She will also look for opportunities to help students recognize the need to adapt their thinking and self-regulate.

Mrs. Reynolds notices Ely erasing her answer and asks her why she is doing this. Ely responds, "I got different answers than Lennox, so I'm probably wrong."

Mrs. Reynolds says, "Just because our answers look different doesn't mean they're incorrect." She reminds Ely about the situation of equivalent forms giving the same value. "Even though the representations looked different, they were the same. Could that be the case here?"

Lennox and Ely review their numbers. Mrs. Reynolds adds, "When your answer differs from someone else's, pause and think why before assuming your answers are wrong. That's called 'self-regulation.' Sometimes different answers mean something is incorrect, or there's a small mistake to fix. Other times, there can be more than one right answer. Your job as mathematicians is to figure that out and adjust your work if needed."

To support student thinking and provide scaffolding, Mrs. Reynolds displays the following information:

40		
400		
4,000		
40,000		
400,000		

She asks, "What do you notice? Write down at least three things." Most students identify the pattern within the zeros.

Lennox doesn't see any patterns, so Mrs. Reynolds asks, "What do you notice when you look at all the numbers?"

Lennox replies, "Fours and zeros."

Mrs. Reynolds then asks, "What do you notice about the zeros?"

Ely chimes in, "The zeros are growing."

Mrs. Reynolds acknowledges Ely's enthusiasm but reminds her, "It's important to be patient and listen to others' ideas. This helps us understand different perspectives, as I'm doing with Lennox. You can also listen and ask questions to help your classmates see things differently."

Turning back to Lennox, Mrs. Reynolds says, "Ely shared her perspective. What do you think she means?" Lennox nods and restates the idea in her own words.

Mrs. Reynolds then shows numbers from the second column and asks, "What do you notice now?" Students think independently before discussing patterns and equality with a neighbor and the class.

40	4×10	
400	$4 \times 10 \times 10$	
4,000	$4 \times 10 \times 10 \times 10$	
40,000	$4 \times 10 \times 10 \times 10 \times 10$	
400,000	$4 \times 10 \times 10 \times 10 \times 10 \times 10$	

As each new idea arises, Mrs. Reynolds asks, "Can you see that? Does that happen anywhere else?" As students present new "notices," Mrs. Reynolds encourages students to come to the front and asks students to "show us where you see that" or asks other students to state another student's thinking using different words. All of this allows students to attend to the structure, patterns, and equivalence.

Finally, Mrs. Reynolds displays the third column of the table and asks, "What do you notice? What do you wonder?"

(Continued)

(Continued)

40	4 × 10	4 × 10^1
400	4 × 10 × 10	4 × 10^2
4,000	4 × 10 × 10 × 10	4 × 10^3
40,000	4 × 10 × 10 × 10 × 10	4 × 10^4
400,000	4 × 10 × 10 × 10 × 10 × 10	4 × 10^5

Students excitedly launch into things they notice. Mrs. Reynolds reminds them that not everyone notices things right away and not everyone sees the same thing, so it is important to have some think time before they share.

Students share connections related to patterns in zero, multiplication by 10, and the base number of 4. Mrs. Reynolds asks Lorynn to share a notice or wonder. Lorynn says, "I wonder if the numbers in each row are all equal," launching the class into another exploration of patterns, structure, and equivalence in a mathematical context while fostering curiosity and deeper thinking. ●

In the discussion, the class explores how powers of 10 affect place value, observing that digits to the left are 10 times larger and digits to the right are $\frac{1}{10}$ as large. They extend this understanding to tenths, hundredths, and thousandths, noting that the digits shift left when multiplying by 10 and right when multiplying by $\frac{1}{10}$.

Mrs. Reynolds cultivates **self-regulation** and **adaptability** by fostering metacognitive awareness, structured discussions, and a supportive learning environment. She teaches students to pause and analyze their thinking, as seen when she encourages Ely to reconsider her answer rather than assume it is incorrect. By tailoring questions to guide Lennox's understanding, she scaffolds adaptability, ensuring students adjust their reasoning rather than passively receiving information. Through collaborative discussions, she prompts students to restate ideas, identify patterns, and engage in peer-supported learning, reinforcing cognitive flexibility. Additionally, she creates a patient and inclusive classroom culture, allowing students time to process, reducing anxiety, and encouraging thoughtful participation. These strategies

CHAPTER 7 • LEVERAGING ADAPTABILITY TO FIND AND USE STRUCTURE

help students independently regulate their learning, adapt to new insights, and develop resilience in problem-solving.

Mr. Maier and His High School Algebra Class

Mr. Maier starts a new algebra unit on graphing functions and identifying key features (HSF-IF.C.7). For the introduction, students examine a projected image in the Which One Doesn't Belong? activity, analyzing a set of graphs to identify the odd one out (Figure 7.6). After thinking individually, they discuss their ideas with a partner, focusing on shifts, asymptotes, and rates of change.

Figure 7.6 • *Functions: Which One Doesn't Belong?*

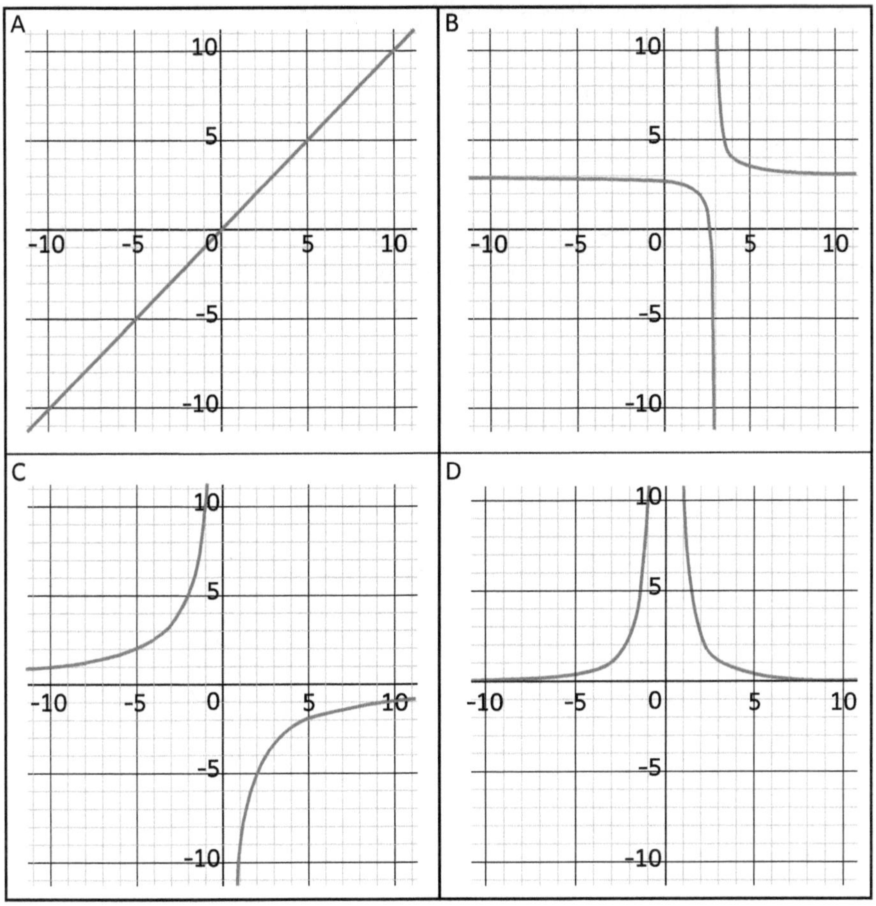

Following partner discussions, Mr. Maier leads a whole class conversation, inviting students to share their thoughts. He pauses after each response, allowing time for students to reflect on the graphs and responses before moving on to the next student.

Keston: Graph A doesn't belong because it's a straight line.

Jaida: Graph B doesn't belong because it spans three quadrants.

Kendal: Graph C doesn't belong because there's nothing in quadrant I.

Kieryn: Graph D doesn't belong because it looks like a volcano.

After the whole class discussion, Mr. Maier says, "You started with one idea about 'which one doesn't belong' and a reason. How have your ideas grown and changed after our discussion?" He provides quiet think time and then encourages students to share with a shoulder partner.

He transitions to the next activity saying, "In this activity, sometimes you took a step back to see the entire graph, and sometimes you focused on one specific character to highlight the differences. You'll want to remember both ideas as you engage in the next activity." After discussing the warm-up activity, Mr. Maier's students explore various inverse variation and rational functions in the form of $f(x) = \frac{a}{x}$ (Table 7.8). For each function, students are instructed to describe how the transformations can be seen in the structure of the equation and how this affects the graph.

Table 7.8 • *Inverse Variation and Rational Functions*

$f(x) = \frac{1}{x+2}$	$f(x) = \frac{1}{x-4}$	$f(x) = \frac{2}{x}$	$f(x) = \frac{-5}{x}$	$f(x) = \frac{6}{x}$
$f(x) = \frac{1}{x-6}$	$f(x) = \frac{-1}{x+3}$	$f(x) = \frac{1}{4x}$	$f(x) = \frac{-2}{x}$	$f(x) = \frac{2}{x+7}$

Emmett and Izzie disagree on the function $f(x) = \frac{2}{x+7}$. Emmett says the function will behave more like the function $f(x) = \frac{1}{x+2}$; Izzie believes the function will behave more like the function $f(x) = \frac{2}{x}$. Mr. Maier notices that both student voices are elevated. Both students seem to have dismissed the other's idea. Mr. Maier asks questions that cause the students to focus on the structure of the function and predict how each aspect of the function will impact the graph.

Soon, Izzie begins erasing her work before the conversation concludes. Mr. Maier asks, "Why are you erasing?" Izzie acknowledges that Emmett's answer is correct and assumes her answer must be wrong. Mr. Maier encourages both students to put down their pencils. After the students converse about their ideas and realize that both students were correct in their thinking, Mr. Maier says, "I appreciate that you are being assertive in sharing your ideas, but it is also important to regulate your emotions so you can have a productive conversation." He continues, "Sometimes in mathematics, there are many correct answers. It's important that we refrain from quickly erasing our work before we've had time to fully investigate all ideas. Even if you are wrong, some of your work will help you get to something that is correct." He asks, "What might you do next time you encounter a similar situation?"

Emmett says, "I can remind myself and my partner that we might both be correct or parts of what we have might be correct."

Izzie says, "I can put my pencil down so I don't erase so quickly, and I can tell myself that it doesn't help when I'm frustrated."

Emmett replies, "Yeah, that's why I just stopped talking about it. It was getting frustrating."

Mr. Maier asks, "What would be more productive than simply shutting down?"

"I can pause and take a deep breath. Then I can suggest that maybe we can focus on one idea at a time," says Emmett. ●

In the first activity, students adapted their thinking to internalize and understand another's viewpoint. The activity scaffolded vocabulary while allowing students to quickly see differences in structure. Mr. Maier could have stopped the activity after students discussed the functions, but he leveraged the student's **adaptability** and see another's perspective with a quick reflection.

In the second activity, students sorted inverse and rational functions, attending to the structure of the equations. Mr. Maier knew Izzie and Emmett were both correct in their thinking. He could have pointed that out to the students rather than having them see that for themselves. Instead, he focused on how to engage in conversation that honored each other's thinking. He asked questions that attended to the structure of the functions, helping the students adapt and intake additional input. He also prompted **self-regulation**. When Izzie began erasing her correct answer, Mr. Maier suggested that she put her writing utensil down to resist the urge to erase. He encouraged a productive conversation after recognizing students' frustration. This allowed students to see another's perspective as well.

Reflection

In both narratives, teachers attended to the structure within mathematics. They connected prior knowledge to new concepts, zoomed in with laser focus to highlight big mathematical ideas and connections, encouraged students to step back for a different perspective, and engaged students in rich conversation that supported visualization and conversation.

Attention to MP7, like the other standards for mathematical practice, shifts emphasis from stopping with the correct solutions to understanding big ideas within mathematics. Opportunities for students to "look for and make use of structure" may arise spontaneously, relying on what students notice in the moment. Teachers need to know how to respond when these opportunities arise. With practice, teachers can strategically provide tasks that attend to structure and pattern while posing purposeful questions that promote these connections.

Designing lessons that activate prior knowledge and support social-emotional competencies further support student's ability to discern patterns and structure. While this chapter addresses adaptability and self-regulation, consider how other social-emotional competencies, like curiosity, would support MP7.

SUMMARY

This chapter examines how teachers use MP7 and social-emotional competencies to enhance lessons. MP7 involves recognizing and making sense of mathematical structure, helping students apply patterns and consider different perspectives. Encourage students to notice mathematical structures through activities like What Do You Notice? or Which One Doesn't Belong?. These routines highlight concepts such as place value, decimal placement, and graph features. Incorporating intrapersonal and interpersonal skills like adaptability and self-regulation is crucial. Students must manage frustration and consider various perspectives when facing challenging problems. For example, Mrs. Reynolds's fifth graders explored patterns in multiplication by powers of 10, focusing on place value and decimals while practicing self-regulation. Mr. Maier's algebra students studied inverse variation and rational functions, emphasizing structure in mathematics and management of emotions. Attention to mathematical structure fosters deeper understanding, effective problem-solving, flexible thinking, and efficient computation. By using MP7 and relevant standards, teachers guide students to focus on mathematical structure and integrate essential social-emotional skills.

> **Questions to Think About**
>
> 1. How can you present an unfamiliar topic so students recognize patterns or structure in a meaningful way (e.g., Instead of telling students about the connection between exponents in power of 10 and the number of zeros, present information and have students generate the patterns and structure they notice.)? What will you do to make students step back and see a different perspective?
>
> 2. In what ways can you support self-regulation in an upcoming lesson?
>
> 3. In what ways can you support adaptability in an upcoming lesson?
>
> 4. Think of a specific mathematics lesson that would incorporate MP7. What other intrapersonal or interpersonal skills would enhance the lesson?
>
> 5. How do you ensure active integration of MP7 and social-emotional competencies into your mathematics lessons?

Actions to Take

1. When you observe students demonstrating an intrapersonal or interpersonal competency, name it, describe it, reinforce it, and affirm it. For example, "You are using adaptability as you listen to your classmates' descriptions of what they see. When we step back and observe mathematics from a different perspective, it _____. Adaptability is important in our classroom and in other aspects of our world. Keep being adaptable!"

2. Write the following on a sticky note:
   - How will I support self-regulation in an upcoming lesson?
   - How will I support adaptability in an upcoming lesson?

   Post this sticky note wherever you make lesson plans to remind yourself to incorporate these social-emotional learning competencies into an upcoming lesson.

3. Consider ways to get students to attend to the patterns and structure of mathematics and another student's perspective/thinking/reasoning.

4. To practice adaptability, ask students to step back and see another's perspective or ideas. Have them share how they processed the new information and adapted their thinking. Ask for a time that they had to do this within the mathematics classroom and a time they adapted their thinking in real life.

CHAPTER 8

INCREASING PERSEVERANCE BY EXPLORING REPEATED REASONING

THINK ABOUT A TIME WHEN YOU WERE INTRODUCED to a rule or procedure in math class that you didn't really understand. You followed the steps although you weren't sure of the "why" of what you were doing—but it just worked. You may be familiar with traditional mathematics textbooks that begin by presenting a "rule" or procedure, followed by examples of how to apply it to specific problem types, and then problems to practice. For some students, this makes mathematics class feel like an endless list of hints, tricks, and shortcuts. This approach of directly teaching the algorithm or shortcut prior to exploration diminishes the cognitive demand of the task and hinders students' abilities to connect concepts to procedures (National Council of Teachers of Mathematics [NCTM], 2017).

In the real world, mathematicians look for regularity in reasoning when confronted with new problems or developing new theories (Goldenberg et al., 2017). Rather than solving single problems, Mathematical Practice 8 (MP8) involves investigating general problem types and draws on reasoning to understand the concepts. Students apply reasoning to repeated calculations to create a general method, rule, algorithm, formula, or shortcut. For example, students might know their doubles in addition from memory (2 + 2, 3 + 3, 4 + 4, etc.), and they then recognize that 2 + 3 can be thought of as 2 + 2 + 1 and 3 + 4 can be thought of as 3 + 3 + 1. They then generate a "doubles plus one" rule for all problems like this.

> **LOOK FOR AND EXPRESS REGULARITY IN REPEATED REASONING**
>
> *Mathematically proficient students notice if calculations are repeated and look both for general methods and for shortcuts. Upper elementary students might notice when dividing 25 by 11 that they are repeating the same calculations over and over again, and conclude they have a repeating decimal. By paying attention to the calculation of slope as they repeatedly check whether points are on the line through (1, 2) with slope 3, middle school students might abstract the equation. $(y - 2) / (x - 1) = 3$. Noticing the regularity in the way terms cancel when expanding $(x - 1)(x + 1)$, $(x - 1)(x^2 + x + 1)$, and $(x - 1)(x^3 + x^2 + x + 1)$ might lead them to the general formula for the sum of a geometric series. As they work to solve a problem, mathematically proficient students maintain oversight of the process, while attending to the details. They continually evaluate the reasonableness of their intermediate results. (CCSSM, 2010)*

While MP7 and MP8 both involve work with patterns and structures in mathematics, there are important distinctions to be made when comparing them. MP7 focuses on utilizing prior knowledge to recognize and use aspects of a mathematical structure to solve a problem. MP8 involves looking for patterns or repeated calculations to create a rule, algorithm, or shortcut. In many ways, MP7 and MP8 work in tandem with one another "empowering learners to uncover the underlying structure of problems, and to prove—and even create—methods and formulas for themselves" (Alvaradous, 2023).

The processes involved in MP8 open the door to both leveraging and building social-emotional competencies. While MP7 involves identifying an arithmetic property, an expression, a formula, or other structures to apply to solve the problem, MP8 is about generating that structure in the first place. In this way, MP8 fosters curiosity and creative thinking. Additionally, "When facing hard problems, mathematicians naturally turn to experiments, looking for regularities that can lead to a theory. This process is not natural to beginners—it must be learned—but becoming proficient in mathematics requires this way of thinking" (Goldenberg et al., 2017, p. 512). This type of engagement also promotes perseverance. Engaging in MP8 allows students to recognize they can think like a mathematician. They feel empowered when they can prove or discover a method. Examining a variety of mathematical situations to identify patterns, search for repetition, generalize patterns, and create shortcuts independently and collaboratively encourages teamwork. In this chapter, we will explore how intrapersonal skills like **perseverance**, **curiosity**, and **creative thinking** along with interpersonal skills such as **teamwork** can be developed through engaging MP8.

MERGING CONTENT STANDARDS, MATHEMATICAL PRACTICES, AND SOCIAL-EMOTIONAL COMPETENCIES

When planning to embed MP8 into the mathematics classroom, it is helpful to consider the decision points that teachers transition through as they select the content standards and goals of the lesson, look for the natural connections to MP8, and identify the intrapersonal and interpersonal skills that are inherently part of the mathematical practice. Leading back to the planning phase of selecting tasks for students to engage in, deciding how they will engage with the mathematics in a way that promotes students' development of shortcuts or generalizable patterns across a variety of mathematical problems.

Mathematical Content Standard and Corresponding Mathematics Goal

Start by asking, **"What is the mathematics goal of this lesson?"** Consider this first-grade standard: "Add within 100, including adding a two-digit number and a one-digit number, and adding a two-digit number and a multiple of 10, using concrete models or drawings and strategies based on place value, properties of operations, and/or the relationship between addition and subtraction; relate the strategy to a written method and explain the reasoning used. Understand that in adding two-digit numbers, one adds tens and tens, ones and ones; and sometimes it is necessary to compose a ten" (1.NBT.C.4, CCSSM, 2010). The primary goal of a lesson aligned to this standard is to add within 100, but to do so in a way that is built on conceptual understandings, such as recognizing patterns in place value and the relationship between addition and subtraction.

Consider the eighth-grade content standard: "Use similar triangles to explain why the slope m is the same between any two distinct points on a non-vertical line in the coordinate plane; derive the equation $y = mx$ for a line through the origin and the equation $y = mx + b$ for a line intercepting the vertical axis at b" (8.EE.B.6, CCSSM, 2010). This standard showcases the emphasis on explanations and deriving equations, as opposed to giving students the equation or formula as described earlier in this chapter. A goal of a lesson aligned to this standard would be for students to examine data to notice repetitions and relationships, leading to the equation.

Mathematical Practice

Now reexamine the language to recognize ways it aligns to the mathematical practices. Ask, **"Which mathematical practice enhances understanding of this content standard?"** Many mathematics standards use verbs such as *find*, *identify*, *derive*, and *generate*. MP8 is a process not an object (Goldenberg et al., 2017), so these words prompt us to recognize the content standard is aligned to MP8 because it provides students with opportunities to generate their own ideas about the relationships, explore patterns in a variety of mathematical computations, and look for overall generalizable structures. This leads to the development of an algorithm that can be replicated in more complex situations in the future. Consider the following content standards:

- Identify arithmetic patterns (including patterns in the addition table or multiplication table) and explain them using properties of operations. For example, observe that 4 times a number is always even, and explain why 4 times a number can be decomposed into two equal addends. (3.OA.D.9, CCSSM, 2010)
- Observe using graphs and tables that a quantity increasing exponentially eventually exceeds a quantity increasing linearly, quadratically, or (more generally) as a polynomial function. (HSF.LE.A.3, CCSSM, 2010)

Specific terms, such as *identify*, *observe*, and *explain*, portray that there are options, like lists of strategies or models, that are aimed at generalizing a concept by exploring different problems. In the kindergarten standard, students are taking what they know about addition and applying it to find an efficient way to determine all the ways to "make 10." This strategy of "make 10" can then be applied to other concepts within kindergarten standards and beyond. By observing patterns over a set of addition problems within 10, students develop the "make 10" strategy. Care must be given to planning how students engage with the content and MP8, as we strive to develop social-emotional competencies within students.

Social-Emotional Competencies

Next ask, **"What intrapersonal and interpersonal skills are inherent, are needed, and can be further developed for students engaging in MP8?"** The integration of the intrapersonal and interpersonal skills with MP8 requires that we are mindful of the connections and purposeful in their implementation, ensuring that these skills are not seen as separate from the lesson content, but rather as inherent aspects of learning. Students may not possess these personal and social skills for various reasons. Therefore, we should explicitly define, teach, and support these skills in a variety of ways during mathematics lessons to ensure their development. There are several intrapersonal and interpersonal skills that naturally correspond to the mathematical tasks students are asked to complete when they engage with MP8.

Intrapersonal Skills

MP8 focuses on students noticing repeated calculations, shortcuts, patterns, and general rules for mathematics across multiple problems, providing an opportunity to develop several intrapersonal skills. Finding regularities across multiple problems has a puzzle like feel. Students' intrinsic interest and curiosity will be demonstrated through their ability to inquisitively find repeated reasoning and relationships across the content. Wondering about mathematical relationships and connections can be modeled and reinforced through searching for relationships, patterns, shortcuts, and connections among items. This promotes **curiosity** and naturally lends itself to **creative thinking**. Curiosity promotes wonder, and creative thinking moves wonder into action, by seeing problems in new ways to generate solutions. Students must consistently push forward when looking for regularity across multiple problems. They persist in finding a general rule or shortcut present within the content, and do not easily give up. Students show **perseverance** as they stay committed to solving the problem even when it takes time or becomes challenging. They grapple with relationships and connections across the content they are examining and utilize the power of "yet" to persist and continue toward finding the answer.

Interpersonal Skills

MP8 calls on students to take the initiative to look for regularity and repetition in their work. Collaborations with peers and teachers expand what they notice and supports students in looking at things from a different perspective and is an opportunity to embed interpersonal skills. MP8 states that students will "maintain oversight of the process, while attending to the details." This is more manageable if they work with others, and in this way, MP8 provides an avenue to support **teamwork**. Students can share results from mathematics learning experiences to collectively produce a shortcut or general rule. More ideas become available when working as a team. Through developing and verbalizing a generalization, pattern, or efficient shortcut, students practice effective communication. Working as a team to stay focused and committed to arriving at a solution also supports the development of perseverance.

Instructional Structures and Engagement Strategies

As we shift toward planning, ask, **"With an eye on our mathematics goal, how will I support social-emotional development as I engage learners in MP8: Look for and express regularity in repeated reasoning? What structures, strategies, methods, and/or tools can I use?"**

In the classroom we can encourage curiosity *and* creative thinking by creating a safe environment for students to feel comfortable taking risks and exploring through the process for learning over finding the "right" answer. We can promote creative thinking by giving ample time to process, explore, and "arrive" at the answer with creative and big ideas.

Fostering Curiosity

Curiosity is a strong desire to know or learn something for its own sake. Research indicates a child's curiosity impacts their ability to grasp basic mathematics and reading concepts (Shah et al., 2018). If harnessed, curiosity can drive learning and achievement as students are naturally curious and wonder about many things in life. We can foster curiosity through intentionally creating an environment that emphasizes and supports healthy questioning and trial and error; focusing on learning through a process rather than feeling pressured to produce the "right" answers. As discussed in the introduction, we can support curiosity by recognizing behaviors we wish to reinforce.

> **Competency Builder 8.1**
>
> *Curiosity-Based Questions and Comments*
>
> Model curiosity for students by sharing your thought process out loud.
> Ask yourself curiosity-based questions and then present them to students.

CHAPTER 8 • INCREASING PERSEVERANCE BY EXPLORING REPEATED REASONING

Acknowledge when you see students using strategies to make sense of patterns that they are curious about. For example, discuss the effectiveness of using a trial-and-error strategy. Prompt curiosity by asking questions and presenting challenges. For example,

- I wonder what patterns I can find.
- I wonder what might be a new way to look at this.
- I wonder how ____ is related to _____.
- I'm curious whether we can find a different approach.
- I'm curious whether we can find a way to be more efficient or find a shortcut to this problem.

These types of questions and comments amplify the skill, draw positive attention to curiosity, and encourage students to implore more of it as it is modeled and reinforced in the classroom. In addition, the way we frame mathematical tasks for students also impacts the approach they take.

Teaching MP8 Explicitly

As stated earlier, MP8 "is not natural to beginners—it must be learned—but becoming proficient in mathematics requires this way of thinking" (Goldenberg et al., 2017, p. 512). We enable students to build these critical thinking skills by providing opportunities for them to explore, explicitly teaching them how to persevere using strategies such as self-talk and asking themselves questions.

Competency Builder 8.2
How to Look for and Make Use of Structure

The following process is recommended for teaching students how to engage in MP8 (Saucedo, 2024). Tell students that there are several steps they can take to "look for and make use of structure." Tell them this will require them to use critical thinking. Ask them what critical thinking means to them. This will also require perseverance. Ask them to notice what thoughts and behaviors help them (or hinder them) from persevering through these steps.

1. Identify the type of problem.
2. Identify what operations within this problem you already know how to do.

(Continued)

(Continued)

3. Identify what you do repeatedly when solving this type of problem.

4. Identify which strategies, rules, or patterns you could use to solve this problem.

5. Identify other shortcuts you could use based on the patterns you noticed.

6. Explain how you can repeat this thinking when solving similar problems.

After engaging students in this process, ask them to share successes and challenges they encountered as they worked to persevere through the task. Discuss strategies for overcoming challenges.

Using Children's Literature

Children's literature is a great source for problem-solving and reasoning tasks. Both literature and mathematics often employ repeated reasoning in several ways. Many stories follow a cyclical or repetitive structure, key themes or messages are often reiterated, and this creates familiarity in the same way that this repeated reasoning in mathematical problems supports students in generalizing reasoning. When students encounter similar problems or concepts repeatedly, they recognize patterns and develop confidence in their ability to identify and tackle challenges. This familiarity motivates them to persevere even when faced with new difficulties. For this reason, the literature experiences shared here support both the development of social-emotional competencies and MP8.

Competency Builder 8.3
When Sophie Thinks She Can't

When Sophie Thinks She Can't by Molly Bang is a book about a girl and her friends trying to form as many different rectangles with 12 tiles as they can (Figure 8.1). When she can't figure it out, her teacher says, "You can't figure it out *yet*," which inspires her to keep trying. We find mathematical problems, the modeling of reasoning processes, and support for social-emotional learning all in one book! This book addresses both perseverance and teamwork. Begin by reading the book aloud. Ask students to share what they noticed Sophie learned in this activity (in addition to mathematics). Define perseverance. Ask students how teamwork

CHAPTER 8 • INCREASING PERSEVERANCE BY EXPLORING REPEATED REASONING

contributed to Sophie's success. She learns to trust her peers. They encourage her to keep going and offer her support when she feels frustrated. Sophie and her classmates learn that teamwork is helping each other through challenges by asking for help and listening to one another. In the end, the celebration is a collective one. Explain how these are very important competencies in mathematics. Students will now have a chance to demonstrate their own perseverance!

Figure 8.1 • *Creating Rectangles*

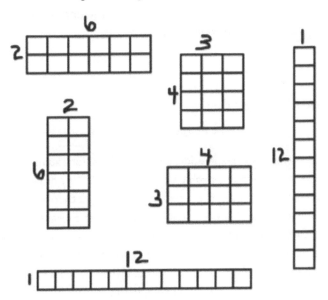

Provide students with 24 squares. Ask them to explore making rectangles with the 24 tiles using different dimensions. Have them summarize this by generalizing with the area formula. In a class discussion, have students reflect on and share the ways they persevered and demonstrated self-efficacy. When responding to students who are struggling and claim that they can't do something, practice responding like the teacher in this book with the statement, "You can't figure it out *yet*." This language references the work of Dweck (2006), where she promotes the notion of fixed and growth mindsets, using the "yet" statement to encourage students to persevere.

Using MP8 to Meaningfully Teach Properties in Mathematics

MP8 calls on students to notice patterns, generalize them, and recognize regularities in mathematics, making it a natural fit with teaching properties from many different branches of mathematics in a meaningful way. Some examples include properties of arithmetic, equality, algebraic or geometric transformation, and exponents and logarithms. You can enhance creative thinking and adaptability by encouraging students to explore and discover mathematical properties themselves, rather than simply presenting properties to them. When students generate properties on their own, and move beyond rote memorization, they internalize the concepts more deeply. They learn how to approach problems, make conjectures, test them, and revise their thinking. This fosters adaptability because they learn how to be flexible in their thinking and adjust based on what works and what doesn't. This gives them a sense of ownership over their thinking. Competencies Builders 8.4 and 8.5 illustrate how this can be done.

Competency Builder 8.4
Introducing the Commutative Property

Present students with two simple addition expressions, for example 4 + 5 and 5 + 4. Ask students if these two problems are the same. Why or why not? Provide students with counters and ask them to demonstrate whether they are the same or not. Then, write 4 + 5 = 5 + 4. Present another pair of expressions, such as 7 + 3 and 3 + 7. Again, ask if the expressions are the same and tell students to show or explain how they know. Show students how to write this as 7 + 3 = 3 + 7. After exploring several examples, ask students if they notice a pattern. Tell them to use this pattern to create their own example to share with a partner. They should exchange examples and try them out. As a class, discuss what they notice about the results of the problems. Ask if it matters if we change the order of the numbers in an addition problem like this. Introduce this as the commutative property that states the order of two numbers that are added (or multiplied) does not change the result. Explain that creative thinking involves adapting one's thinking. Ask students how they had to adapt their thinking about addition as they explored the commutative property in this activity. They may have noticed that they had to adapt their thinking when they used counters to build the expressions because they had to be very aware of which number came first. Ask them how adapting their thinking helped them to be creative. They may notice that being able to adapt their thinking enabled them to take the examples they worked with and use the pattern to create their own example. Ask them to think about how they use this type of thinking in other aspects of their life.

CHAPTER 8 • INCREASING PERSEVERANCE BY EXPLORING REPEATED REASONING

> **Competency Builder 8.5**
> *Introducing Properties of Exponents*
>
> Like Competency Builder 8.4, this activity invites students to investigate several examples to look for a pattern and make a generalization. Consider the exponent property product of a power. Present students with an expression and ask them what it means. For example, present students with the expression $5^2 \times 5^3$. Ask students what this expression means. They should recognize that this is 5 multiplied by itself 2 times and 5 multiplied by itself 3 times. Ask students to write this out to demonstrate what they just said. This would look like $5 \times 5 \times 5 \times 5 \times 5$. Ask them what they notice. They should notice this is 5 multiplied by itself 5 times. Ask students to write this in simplified form. Have them do several more problems like this as illustrated in Figure 8.2.
>
> **Figure 8.2** • *Product of a Power Examples*
>
> $5^2 \times 5^3 = 5 \times 5 \times 5 \times 5 \times 5 = 5^5$
> $4^4 \times 4^5 = 4 \times 4 \times 4 \times 4 \times 4 \times 4 \times 4 \times 4 \times 4 = 4^9$
> $2^8 \times 2^3 = 2 \times 2 \times 2 \times 2 \times 2 \times 2 \times 2 \times 2 \times 2 \times 2 \times 2 = 2^{11}$
> $9^2 \times 9^6 = 9 \times 9 \times 9 \times 9 \times 9 \times 9 \times 9 \times 9 = 9^8$
>
> Ask the students to look for a pattern. They should notice that the exponent in the simplified expression is the sum of the exponent with the same bases. Ask them to create a "rule" that would generalize this pattern. One way they could do this would be to write it as $y^a \times y^b = y^{a+b}$. Introduce other exponent properties in the same fashion. Remind students that creative thinking requires them to adapt their thinking. Ask students how they had to adapt their thinking about exponents as they explored the properties. They should notice that they had to rewrite expressions in different ways to make sense of how the property works. Ask them how adapting their thinking helped them to create a rule to generalize the property. They should recognize that adapting their thinking helped them to take the examples they worked with and apply creative thinking to write a rule to represent it. Ask them to think about how they use this type of thinking in other aspects of their life.

Using Games

Games and routines provide excellent opportunities to look for and express regularity in repeated reasoning. Guess My Rule is a game used across K–12.

Competency Builder 8.6
Guess My Rule Game

Introduce the game to students by telling them this game will involve teamwork. Ask them what teamwork requires. In this game there is a leader, often called the "rule keeper," who makes up a rule (or can be supplied one from a deck of "rule" cards) that gives exactly one output for each input. Players take turns going round-robin giving a number for the input, and the rule keeper gives the output. This continues until a player wants to guess the rule. To do so, they give both an input and an output that they think would work. If they are right, they get to guess the rule. If they guess correctly, they become the new rule keeper.

For example, see if you can guess my rule. Student one suggests 5 as the input and the rule keeper says the output is 8. Student two gives 10 as the input and the rule keeper says the output is 13. Student three says she wants to guess the rule and says, "My input is 1 and output is 4." To which the rule keeper responds telling her she is right and asks for the rule. She says the rule is some number plus 3. We record the inputs, outputs, and rule as we play (Figure 8.3).

Figure 8.3 • *Input-Output Table for the Rule n+3*

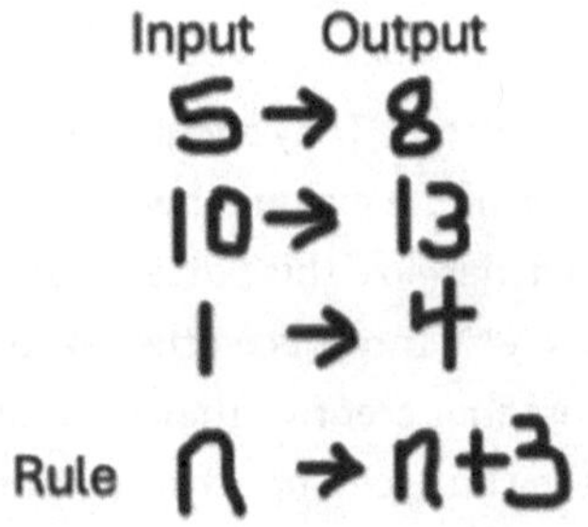

$$\begin{array}{cc} \text{Input} & \text{Output} \\ 5 & \rightarrow 8 \\ 10 & \rightarrow 13 \\ 1 & \rightarrow 4 \end{array}$$

Rule $n \rightarrow n+3$

Generating a list of inputs and outputs in this way allows students to engage with MP8 using the "guess-check-generalize" approach (Goldenberg et al., 2017). As inputs and outputs are added to the table, students begin to do their own guessing at what the output will be, checking based on what the rule keeper tells them. When they find their reasoning checks, they use the repeated reasoning they employed in the game to generalize the examples into a rule. This type of thinking also supports students in learning how to "maintain oversight of the process, while attending to the details." Depending on the grade level and the

> content studied, the functions used for this activity vary and start simple like the example provided and grow in complexity. There are also options for playing Guess My Rule through online platforms such as Desmos. The game-like features of this task, asking students to guess a rule each time they are given an input and output, stimulates curiosity and creative thinking. This creative thinking can be harnessed and shifted to critical thinking as they leverage repeated reasoning to generate a rule. The game format and interaction with peers encourages perseverance and teamwork.

Assessing Interpersonal and Intrapersonal Skills

Next we ask, **"How will I assess students' progress toward the mathematics goal of this lesson, their engagement in the mathematical practice standard, and their ability to use and continue to develop intrapersonal and interpersonal skills? How will I provide feedback?"** These questions can be answered using various assessment strategies, both formative and summative. A whole class (Table 8.1) or individual student observation tool (Table 8.2) serves as another means of informal assessment, and includes mathematics goals, engagement with components of MP8, and development of social-emotional competencies aligned to lesson. Self-assessments are another effective instructional and assessment strategy, allowing students to determine their self-efficacy relating to specific social-emotional competencies. The following self-assessment checklist could be used on individual lessons, or over time, to help students self-identify their social-emotional strengths and weaknesses (Table 8.3). Use the following prompts during or after a lesson to guide student reflection. These same prompts can be given to students to help them reflect on their own learning.

- How was curiosity a part of the lesson?
- In what ways did you use creative thinking to find and generalize a pattern?
- What does perseverance mean to you and how did you demonstrate it today?
- How did teamwork enable you to be successful today? What role did you play?

Table 8.1 • *Whole Class Observation Tool*

Name	Mathematics Goal	Engagement in Practice Standard *Notice repetitions*	Engagement in Practice Standard *Generate rule or shortcut*	Intrapersonal Competency *Creativity*	Interpersonal Competency *Teamwork*

Note: Progress will be marked using 0–No evidence, 1–Little evidence, 2–Adequate evidence

online resources — Download this table at https://companion.corwin.com/courses/wellroundedmathstudent

Table 8.2 • *Individual Student Observation Tool*

Name of student:

Mathematics Goal	No Evidence	Little Evidence	Adequate Evidence
Engagement in the Practice	No Evidence	Little Evidence	Adequate Evidence
Notice repetitions			
Generate rule or shortcut			
Social-Emotional Competencies	No Evidence	Little Evidence	Adequate Evidence
Perseverance			

Social-Emotional Competencies	No Evidence	Little Evidence	Adequate Evidence
Teamwork			
Curiosity			

Download this table at https://companion.corwin.com/courses/wellroundedmathstudent

Table 8.3 • *Self-Assessment Checklist*

Social-Emotional Competency	Not Sure	Not Yet	Getting There	Got It
Perseverance				
Creativity				
Curiosity				
Teamwork				
Other skills used:				

Download this table at https://companion.corwin.com/courses/wellroundedmathstudent

These observation, monitoring, and self-assessment tools can help teachers and students determine their progress toward developing social-emotional competencies and mathematical concepts.

LOOKING AT EXEMPLARS IN ACTION

Now that the merging of content standards, mathematical practice standards, and social-emotional competencies has been explored, let's look more closely at an elementary and a secondary example, using the standards and competencies focused on in this chapter.

Mrs. Sharp and Her First-Grade Class

Mrs. Sharp and her first-grade class are continuing to learn about place value as they start adding within 100. Students previously used cubes, base-ten blocks, counters, tape diagrams, and number bonds to model single-digit addition and subtraction and wrote number sentences to connect models to algorithms. Now they are ready to begin focusing on adding two-digit numbers. Mrs. Sharp is focused on having students "Add within 100, specifically focusing on adding a two-digit number and a multiple of 10" (1.NBT.C.4, CCSSM, 2010).

Mrs. Sharp launches the lesson with a Notice and Wonder routine adapted from her curriculum resource. She knows her class loves stickers, so she uses this as a context. Holding up several small bags of stickers she says, "I want to buy some stickers for our class. I already have some stickers in bags for each group, but I want to buy more. At the store, the stickers come in packs of 10. Can you help me figure out how many stickers I should buy? Talk with your table group to answer the questions, 'What do you notice?' and 'What do you wonder?'"

After their notice and wonder discussion, Mrs. Sharp puts the students in groups and explains she is going to give them several "What if . . ." situations. She points out that starting sentences with "What if . . ." encourages creativity and challenges them to do this in everyday situations. She encourages them to look for a pattern in the "What if . . ." scenarios. She instructs them to do two things: First, justify their answer with at least one type of manipulative or picture, and second, write a number sentence to match the problem. She explains they will share their answers, and their answers will all build on one another.

All groups start with 23 stickers in their bag. She asks, "What if I buy one pack of stickers to put in your bag? How many stickers will you have?" Mrs. Sharp monitors students as they work, using an individual student observation tool like Table 8.2. She notices a group lined up counters to represent the stickers in the bag and 10 counters to represent a new pack of stickers and count them one-by-one. She notices another group drew pictures of 10 stickers to model the situation and then counted from 23 by 1 until they reached 33. All groups wrote the number sentence 23 + 10 = 33. She asks, "What if I buy 2 packs of stickers for your group? How many stickers will

CHAPTER 8 • INCREASING PERSEVERANCE BY EXPLORING REPEATED REASONING

you have in your bag?" Some groups used approaches like what they did previously, but she notices a few more groups started counting on from 23 or 33 rather than count all. One group uses base-ten blocks to model the tens and ones in their sticker collection and the two packs of 10 stickers. They counted all the tens (40) and the ones (3) and wrote the number sentence 23 + 20 = 43. She has groups share how they modeled the situation, their strategy, and their answer. She adds this to the table.

She asks, "What if I buy 3 packs of stickers for your group? How many stickers will you have in your bag?" At this point, most groups have abandoned drawing more stickers. More groups move to using base-ten blocks to represent the situation. One group draws a place value chart and makes 2 tallies in the tens and 3 tallies in the ones columns and then makes three more tallies in the tens column. After groups share their model and how they used it, they share they have written 23 + 30 = 53 and Mrs. Sharp records this in the table.

She asks, "What if I buy 4 packs of stickers for your group? How many stickers will you have in your bag?" Groups continue to use base-ten blocks and the place value chart, but she now observes a group draw a horizontal bar and label one part 23 and then write 10, 10, 10, 10 in the other parts (Figure 8.4). They proceed to start at 23 and count by tens.

Figure 8.4 • *Bar Model Representation*

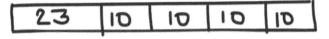

Again, she has them share their model, strategy, and answer. This continues for several more packs of stickers as students experiment with different ways to represent the situation and different strategies for adding multiples of 10. In the end, they have constructed a running record of the addition problems they worked on in the sticker "What if . . ." situation.

(Continued)

(Continued)

23 + 10 = 33

23 + 20 = 43

23 + 30 = 53

23 + 40 = 63

23 + 50 = 73

23 + 60 = 83

Mrs. Sharp prompts them to think about patterns in how they modeled the sticker problem and what all of these "What if . . ." situations had in common. After a minute of think-time she engages them in a discussion.

Sam says, "We noticed that the 23 was the same each time."

Mrs. Sharp asks, "What is the 23, what does it represent in the problem?"

"The stickers we already had," replies Juanita.

"Why do you think it's always the same?" questions Mrs. Sharp.

Yolanda responds with, "We all started with the same number of stickers."

Gavin chimes in saying, "The number we added was 10 bigger each time."

They notice in the running record that the answers went up by one in the tens place.

Mrs. Sharp states, "I wonder what's happening here. What are these numbers and why are they different each time? Talk to you partner and see what you think."

After a brief discussion, Anissa says, "Those are the stickers you are buying."

Mrs. Sharp inquires, "Could I buy 5 stickers? Why is it always 10 more?"

Anissa answers, "Because the stickers come in packs of 10. You could maybe find different packs with 5."

Mrs. Sharp continues, "OK, what else do you notice?"

"They all end in 3, not 0 even though we add 10," observes Marcus.

Mrs. Sharp asks, "Can anyone help us remember what those digits are called?"

Pointing to the digits, Tami says, "The first number is the ones, and the second number is the tens."

Mrs. Sharp asks, "So, why is there a 3 in the ones place?"

"Because we started with 23 which is 2 tens and 3 ones," Duncan replies.

"How does the adding tens show up in our list?" asks Mrs. Sharp

"You can see the tens place is going up 1 each time we add a pack of 10 stickers," says Tami pointing to the list.

Hovena adds, "We used a hundred chart. We circled 23 and then we went down the chart and circled the number below it . . . 23, 33, 43, 53, 63, 73, and 83. The ones digit is always 3 because you start with 23, but the tens digit is 1 bigger each time."

Mrs. Sharp knew this language might lead to a misconception, so she asked, "It is 1 bigger?"

Tami answers, "No, it's really 10 bigger, but the digit is 1 more, like 2, 3, 4, 5, 6, 7, 8."

Mrs. Sharp makes several moves to model curiosity. She uses words like *curious*, *wonder*, and *what if* to instill this in her students. She provides ample time for them to notice, wonder, explore, and explain. She shows perseverance by exploring various ways to model the situation and demonstrating patience in finding the precise words to explain what it means to add 10 repeatedly to a two-digit number.

THE WELL-ROUNDED MATH STUDENT

Mrs. Slaman and Her Eighth-Grade Class

Mrs. Slaman wants her student to learn to "Use similar triangles to explain why the slope m is the same between any two distinct points on a non-vertical line in the coordinate plane" (8.EE.B.6, CCSSM, 2010). As the first lesson in the unit, it will introduce students to the idea of using similar triangles to explain slope by asking them to look for and express regularity in repeated reasoning in the context of a set of stairs.

She launches the lesson with a Notice and Wonder routine using several images of different sets of stairs. After a discussion, she introduces the lesson saying, "I know several people that have fallen up or down the stairs in front of our building. They complain the stairs are too steep. Today, we are going to try to figure out how steep the stairs are so we can compare that to other sets of stairs. Your job is to come up with a way to measure and put a number on how steep the stairs are. We will have 10 minutes outside to collect information, then we will come back to the classroom and share our findings." Students gather paper, pencils, graph paper, rulers, calculators, and some want to bring their phones to take pictures (allowed for this activity).

Students attempt multiple solution paths, measuring the base and height of each stair, lining up rulers end to end to measure the entire length and height of the set of stairs, and measuring the horizontal distance from stair to stair. While students are investigating the stairs, Mrs. Slaman uses the whole class observation tool (Table 8.1). She notes which students are demonstrating creativity and curiosity, she notes students which students are struggling to start (potential lack of these competencies), and she notes students who tried one thing, then just sat down and seemingly gave up (lack of perseverance).

Back in the classroom Mrs. Slaman asks, "Now that we have collected some information about our stairs, what should we examine first?"

Coley says, "We measured all the stairs, so we should see if they are the same." Students share and she records the information (Table 8.4).

Mrs. Slaman says, "Talk to your partner, what do you notice? What do you wonder?"

Vicki says, "The base and the height are almost the same. They are all close to nine inches for the base and seven for the height."

CHAPTER 8 • INCREASING PERSEVERANCE BY EXPLORING REPEATED REASONING

Mrs. Slaman says, "It is natural to find minor errors when we make real-world measurements. For today's purpose, let's use Vickie's suggestion. Using the table of info, work with your partner to determine a way to represent this. How would you show and describe how steep the stairs are in terms of slope?"

Carlos says, "When I hear slope I think of a hill. Maybe we can draw it and show it as a diagonal line like a hill."

"We can use graph paper and use the squares to show the height and base for the stairs," suggests Ryker.

Ryker and Carlos share their idea with the class. Ms. Slaman asks, "What goes on the x-axis? What goes on the y-axis?"

Yanci answers, "It makes sense for the height to go on the y-axis because it's vertical and the base would go on the x-axis because it is horizontal."

Table 8.4 • *Initial Stair Data Collected*

Stair	Base (in)	Height (in)
1	$8\frac{7}{8}$	$7\frac{1}{8}$
2	9	7
3	$9\frac{1}{8}$	$7\frac{1}{8}$
4	9	7
5	$9\frac{1}{8}$	7
6	9	$6\frac{7}{8}$
7	9	7

(Continued)

(Continued)

Students begin graphing the stairs going up 7 and over 9 each time, drawing lines and making a point (Figure 8.5). Mrs. Slaman asks, "If we list these as coordinates, for each stair, what are the coordinates when you are on the ground?"

Figure 8.5 • *Example of Student Work Sample, Graph*

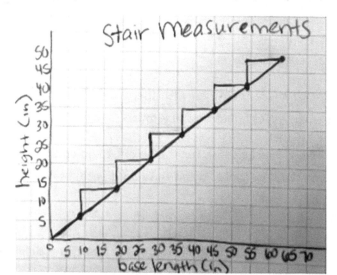

Elizabeth says, "You would be at 0 comma 0."

"What about this point for stair number 2?" Mrs. Slaman asks.

"That one would be 18 comma 14," says Jamarcus.

Mrs. Slaman probes further, "What does that mean in terms of the stairs?"

Nadie answers, "When your foot hits the second step it is 18 feet forward and 14 feet higher than when it started."

Mrs. Slaman suggests, "Let's make a table with the information presented this way so we can see the more steps we take the higher we are" (Table 8.7).

Table 8.5 • Revised Stair Data

Base (in)	Height (in)
9	7
18	14
27	21
36	28
45	35
54	42
63	49

Mrs. Slaman asks students to notice things about the graph. In pairs, students discuss and then share the following: (1) the graph makes a straight line (linear relationship), (2) all the numbers go up by 9 and over 7 each time, (3) all the points are the same distance away from each other, and (4) it looks like a bunch of similar triangles.

Mrs. Slaman asks, "How do you know they are similar triangles?"

A discussion ensued around the patterns in the graph and in the table. Mrs. Slaman continues to foster curiosity and MP8 by asking additional questions such as, "So, what if we had a different flight of stairs, how would we go about finding the slope? What if we had a graph of a line representing the steepness, how would we find the slope?"

Mrs. Slaman engaged her students in critical thinking activities, aimed at generating a rule based on their exploration activity with slope. They overcame the frustration of remaking tables by persevering and communicating with their peers through teamwork. They developed these interpersonal and intrapersonal skills while learning the mathematical concept of slope.

Reflection

Utilizing Notice and Wonder routines, questioning strategies, and facilitating meaningful discussion are a few of the instructional strategies teachers can use to promote students' engagement in MP8 and the development of interpersonal and intrapersonal skills. MP8 focuses on students deriving rules or shortcuts based on their exploration and analysis across multiple problems or situations, looking for consistencies and patterns that develop over time. Students' curiosity can be peaked by working on high-quality mathematical tasks, while teachers implement Notice and Wonder routines with sustained push toward getting students to explore problems from multiple perspectives, looking further into patterns to see consistencies or irregularities, that lead toward shortcuts or rules. This accompanies the strategies we use to develop students' interpersonal skills of creativity. We use questioning strategies to maintain the cognitive demand of tasks and advance student thinking while engaging in Notice and Wonder routines while analyzing sets of problems. Developing perseverance, as it applies to mathematical thinking and beyond the classroom, takes time. We can promote this interpersonal skill by facilitating meaningful discussions, where students are encouraged in sustained critical thinking. Self-assessments of perseverance also play a role in helping students see their physical, emotional, and social reactions to struggle and their ability to persevere through struggle, or not.

SUMMARY

In this chapter, we explored the focus of MP8, looking for consistencies across multiple problems to derive rules or shortcuts that can be applied to other mathematical problems. In doing so, students develop intrapersonal skills of curiosity, creativity, and perseverance, as well as interpersonal skills of goal setting. In both classrooms we explored, teachers facilitated meaningful discussions by engaging students in Notice and Wonder routines and questioning strategies, using observation and monitoring tools to further promote these social-emotional competencies and mathematical concepts. Now we will consider specific questions to think about and actions to take that promote the development of social-emotional competencies that can enhance the teaching and learning of mathematics.

Questions to Think About

1. How can you examine your mathematical content standards for inclusion of MP8?

2. Which of your grade-level content standards require students to notice regularities and repetitions across multiple problems and make connections to and/or generate rules or shortcuts?

3. Examining your curriculum, which types of tasks could students collaboratively engage in to develop a conceptual understanding of a commonly taught rule, trick, or algorithm?

4. Which instructional strategies could you use to encourage creative thinking, curiosity, and perseverance as your students explore repetitive relationships within mathematics?

5. How can you assess your student's social-emotional competencies development while engaged in mathematical learning?

Actions to Take

1. Identify rules, algorithms, patterns, and regularities that students can derive through the examination of repetitive relationships at your grade level.

2. Explore your curriculum resources and/or adapt your mathematical tasks to encourage students to notice repetitions and regularities in mathematics patterns, as opposed to showing or telling students about these rules.

3. Select two to three instructional strategies that promote the competencies that naturally align with MP8, creative thinking, perseverance, and curiosity, and try them in your classroom.

4. Use observation tools, like those provided in this chapter, to monitor students' development of content and practice standards, as well as intrapersonal and interpersonal skills.

CHAPTER 9

FROM AWARENESS TO ACTION

Final Thoughts on Merging Social-Emotional Competencies With Math Practices

THIS BOOK HAS PROVIDED A FRAMEWORK FOR ALIGNING SOCIAL-EMOTIONAL competencies to academic outcomes, illustrating how to dovetail and teach these next generation skills within your math lesson rather than treating them as an add-on. It is our hope that we have effectively shared how you can use what you already do—with small but intentional shifts—to equip your students to thrive in a rapidly evolving world. The competency builders shared in each chapter illustrate how you can creatively incorporate social-emotional competencies into your instruction. This chapter reflects on the important anticipated outcomes of implementing activities that merge content standards, practice standards, and social-emotional competencies, reviews the steps involved in planning for and assessing social-emotional competencies, and provides responses to frequently asked questions that may arise as you implement lessons that merge them.

REFLECTING ON THE OUTCOMES OF IMPLEMENTING COMPETENCY BUILDERS

Several key outcomes are anticipated from implementing competency builders that merge social-emotional competencies with standards for mathematical practice. Tweaking and enhancing what you already do using careful integrative lesson planning will allow you to develop and nurture the whole student. Anticipated outcomes include but are not limited to the following:

- Enhanced learning environments where students trust, feel safe, and have a sense of belonging.
- Strengthened social awareness among students characterized by respectful interactions, perspective taking, and the valuing of different approaches.

- Heightened engagement, which enables students to feel connected to the material resulting in increased knowledge retention.

- Improved adaptability and flexibility to tackle complex math problems.

- Increased resilience demonstrated by an ability to view challenges and mistakes as opportunities for learning rather than obstacles.

- Strengthened emotional regulation marked by skills to manage anxiety and frustration related to learning math.

- Effective collaboration and communication defined by their ability to build relationships, acknowledge varying perspectives, and work effectively with peers.

- Boosted academic performance and strengthened positive attitude toward the subject.

REVIEWING THE PLANNING PROCESS

The purpose of this book is to help you feel more comfortable and confident in highlighting social-emotional skills in your lessons. With thoughtful planning and small changes, we bring intrapersonal and interpersonal skills to the forefront, moving toward a more holistic way of teaching. At the start of the book, we introduced a set of guiding questions, and we modeled the process throughout each chapter to support your planning to make social-emotional competencies a key part of your lessons. The framework begins by identifying a math practice that aligns with your lesson's content goals and the thinking and behavior it requires. Next, we select one or two social-emotional competencies that support this practice and that your students will need for the lesson activities. It's important to be intentional by "naming and framing" these skills. Finally, as you wrap up the lesson, remember to gather evidence to see how your students are developing these social-emotional skills. What are the small steps you can take to make this shift? You can

- Be intentional
- Be purposeful
- Be focused
- Engage in professional development
- Collaborate with colleagues—build a supportive culture

- Start small
- Utilize resources
- Monitor progress
- Reflect and adapt
- Communicate with stakeholders

In summary, integrating social-emotional competencies with math practices enriches mathematics learning experiences and prepares students for future academic and personal success. The more students develop social and emotional skills, the greater their ability to enhance the well-being of their classes, schools, families, and communities. Table 9.1 shares some frequently asked questions and offers advice on how to respond to them as you collaborate with students, colleagues, and caregivers.

Table 9.1 • *Frequently Asked Questions (FAQs) and How to Respond to Them*

Question	Responses
What are social-emotional competencies?	Sometimes SECs are referred to as "soft skills" or "employability skills" (Jones et al., 2021). They are interpersonal and intrapersonal skills that employers are looking for and are essential for successful careers. CASEL (n.d.c), a key organization formed from a large network of parents, educators, and researchers, identified five main competencies: self-awareness, self-management, social awareness, relationship skills, and responsible decision-making.
Why does my child need to develop SECs?	As described in the introduction, youth mental health is a growing concern in the United States. Research indicates that students who develop social-emotional skills are better equipped to combat issues that create anxiety, stress, depression, lack of confidence, and frustration as well as demonstrate improved academic performance. These competencies are tools that help students better understand their thoughts, feelings, and behaviors in stressful situations. All learning is socially and emotionally connected, and social-emotional competencies are essential for preparing students for college, career, and community life.

(Continued)

(Continued)

Question	Responses
Why include teaching about SECs in math class?	There is a close connection between the math practices we want to develop in students and the social-emotional competencies. For example, learning mathematics can sometimes be a frustrating experience, requiring students to persevere, listen to and work with others, and adapt their thinking and behavior to enable them to be more successful. In addition, the stronger students become with these competencies, the more success they can experience inside and outside of the math classroom.
How are SECs integrated into the daily math classroom routines?	We can integrate SECs into the classroom as students solve problems, explain their thinking, listen to others share, discuss ideas, make decisions about what math tools to use, make sense of new ideas, demonstrate perseverance, and more. The SECs are naturally a part of what we do in math but integrating them into instruction means we name them when we use them and help students be aware of how they are developing these skills, so they are visible. Teachers can create a supportive environment in the classroom by teaching social and emotional skills directly, modeling positive behavior, and helping students practice these skills.
Doesn't including instruction focused on SECs take time away from learning math?	Developing SECs in tandem with learning math helps students to develop and strengthen both skills simultaneously. For students who are apprehensive about math, they can leverage the SECs in the lesson to build success in the math classroom.
How are SECs assessed or measured?	SECs develop on a continuum. They are not a "have it" or "don't have it" type of skill, but rather skills that develop and build over time and with repetition. SECs are often assessed informally through observation and by listening for them in practice.
What can families do to support the development of SECs?	We are not born knowing how to manage our emotions, control our behavior, or handle tough situations. These skills are learned over time, and both teachers and families play an important role in building them. Families can support this learning by setting a good example through their words, attitudes, and actions. At home, families can talk about the social-emotional competencies students are learning in school and look for ways to practice them together. This teamwork between teachers and families helps students grow stronger in managing their emotions and behavior.

CHAPTER 9 • FROM AWARENESS TO ACTION

According to a report by Dell Technologies (2017), the job market is changing so rapidly that 85 percent of the jobs that will exist in 2030 have not been invented yet. This underscores the fast pace of technological change and the growing uncertainty about what the future workforce will look like. It points to the importance of preparing for a future that requires constant adaptation and learning, so we are continuously embracing these changes rather than fearing them. It is easy to see why social-emotional learning is so critical for students' future success.

REFERENCES

Abrams, Z. (2023). Kids' mental health is in crisis. Here's what psychologists are doing to help. *Monitor on Psychology, 54*(1), 63–67. https://www.apa.org/monitor/2023/01/trends-improving-youth-mental-health

Acosta, K. (n.d.). *3 act tasks*. https://kristenacosta.com/3-acts/

Aisling, L., Hourigan, M., & McMahon, A. (2013). Early understanding of equality. *Teaching Children Mathematics, 20*(4), 246–252. https://doi.org/10.5951/teacchilmath.20.4.0246

Alvaradous, M. (2023, February 14). *Deep dive: How math practices 7 and 8 power students "lightbulb" moments*. EdReports. https://edreports.org/resources/article/deep-dive-how-math-practices-7-and-8-power-student-lightbulb-moments

Arnold, E. G., Burroughs, E. A., Carlson, M. A., Fulton, E. W., & Wickstrom, M. H. (2021). *Becoming a teacher of mathematics modeling, K-5*. National Council of Teachers of Mathematics.

Ayalon, M., & Even, R. (2016). Factors shaping students' opportunities to engage in argumentative activity. *International Journal of Science and Mathematics Education, 14*(3), 575–601. https://doi.org/10.1007/s10763-014-9584-3

Ayalon, M., & Hershkowitz, R. (2018). Mathematics teachers' attention to potential classroom situations of argumentation. *Journal of Mathematical Behavior, 49*, 163–173.

Bandura, A. (Ed.). (1995). *Self-efficacy in changing societies*. Cambridge University Press. https://doi.org/10.1017/CBO9780511527692

Bandura, A. (2001). Social cognitive theory: An agentic perspective. *Annual Review of Psychology, 52*, 1–26.

Bay-Williams, J. M., & SanGiovanni, J. (2021). *Figuring out fluency in mathematics teaching and learning, grades K-8: Moving beyond basic facts and memorization*. Corwin.

Bay-Williams, J. M., SanGiovanni, J., Walters C. D., & Martinie, S. (2022). *Figuring out fluency – Operations with rational numbers and algebraic equations: A classroom Companion*. Corwin.

Benson, J. (2021). *Improve every lesson plan with SEL*. ASCD.

Bermejo, V., & Diaz, J. J. (2007). The degree of abstraction in solving addition and subtraction problems. *The Spanish Journal of Psychology, 10*(2), 285–293.

Boaler, J. (2019). Developing mathematical mindsets: The need to interact with numbers flexibly and conceptually. *The American Educator, 42*(4), 28–33.

Bonella, L., Carroll, D., Jobe, M., Kaff, M., Lane, J., Martinez, T., McKeeman, L., & Shuman, C. (2020). *Access, engagement, and resilience during Covid-19 remote learning* [White paper]. Kansas State University College of Education. https://coe.kstate.edu/research/documents/KSU-COE-White-Paper-7-2020.pdf

Bouffard, S. (2018). Teaching is an art - and a science. *Learning Forward, 39*(6), 5.

Brodsky, J. (2021, November 22). Why questioning is the ultimate learning skill. *Forbes*. https://www.forbes.com/sites/juliabrodsky/2021/12/29/why-questioning-is-the-ultimate-learning-skill/

Brown, R. (2017). Using collective argumentation to engage students in a primary mathematics classroom. *Mathematics Education Research Journal, 29*, 183–199. https://doi.org/10.1007/s13394-017-0198-2

Burke, K. (2010). *Balanced assessment: From formative to summative*. Solution Tree Press.

Burns, M. (1996). Instructor. *7 musts for using manipulatives*. Scholastic.com. Retrieved May 26, 2024, from https://marilynburnsmath.com/library/MathManipulatives.pdf

Bushweller, M. (2022). How educators view social-emotional learning, in charts. *EdWeek*. https://www.edweek.org/leadership/how-educators-view-social-emotional-learning-in-charts/2022/11

Centers for Disease Control and Prevention. (2021). *Youth risk behavior survey data*. https://www.cdc.gov/yrbs

Centers for Disease Control and Prevention. (2024). *Youth risk behavior survey data summary & trends report: 2013–2023*. U.S. Department of Health and Human Services.

Cirillo, M., Pelesko, J. A., Felton-Koester, M. D., & Rubel, L. (2016). Perspective on modeling in school mathematics. In J. L. Johnson & A. S. Ochs (Eds.), *Mathematical modeling and modeling mathematics* (pp. 1–15). National Council of Teachers of Mathematics.

Collaborative for Academic, Social, and Emotional Learning. (n.d.a). *Fundamentals of SEL*. https://casel.org/fundamentals-of-sel/

Collaborative for Academic, Social, and Emotional Learning. (n.d.b). *Social and Emotional (SEL) key messages*. https://casel.s3.us-east-2.amazonaws.com/B2S-Key-SEL-Messages.pdf

Collaborative for Academic, Social, and Emotional Learning. (n.d.c). *What is the CASEL framework?* https://casel.org/fundamentals-of-sel/what-is-the-casel-framework/

Common Core State Standards Initiative. (2010). *Common core state standards for mathematics*. https://www.thecorestandards.org

Connor, A., Singletary, L. M., Smith, R. C., Wagner, P. A., & Francisco, R. T. (2014). Teacher support for collective argumentation: A framework for examining how teachers support students' engagement in mathematical activities. *Educational Studies in Mathematics, 86*, 401–409. https://doi.org/10.1007/s10649-014-9532-8

Davidson, D. (1991). Children's decision-making examined with an information-board procedure. *Cognitive Development, 6*, 77–90.

Dell Technologies. (2017). *The future of work report*. Author.

Diaz, R. M., & Berk, L. E. (1992). *Private speech: From social interaction to self-regulation*. Lawrence Erlbaum Associates.

Dweck, C. S. (2006). *Mindset: The new psychology of success*. Random House.

Elevated Achievement Group. (2020). *Teach your students to attend to precision*. https://elevatedachievement.com/articles/teach-your-students-to-attend-to-precision/

England, L. (2015, June 22). *Engaging students in three-acts, Part 2*. https://www.nctm.org/Publications/TCM-blog/Blog/Engaging-Students-in-Three-Acts,-Part-2/

Fletcher, G. (2023, February 10). *Packing sugar*. Questioning My Metacognition. https://gfletchy.com/packing-sugar/

Fletcher, G. (n.d.). *3-act task file cabinet*. https://gfletchy.com/3-act-lessons/

Francisco, J. M., & Maher, C. A. (2005). Conditions for promoting reasoning in problem solving: Insights from a longitudinal study. *The Journal of Mathematical Behavior, 24*(3–4), 361–372. https://doi.org/10.1016/j.jmathb.2005.09.001

Garfunkel, S., & Montgomery, M. (Eds). (2019). *GAIMME: Guidelines for assessment and instruction in mathematical modeling education* (2nd ed.). COMAP/SIAM.

REFERENCES

Gibbons, P. (2002). Learning language, learning through language, and learning about language: Developing an integrated curriculum. In P. Gibbons (Ed.), *Scaffolding language, scaffolding learning: Teaching second language, learners in the mainstream classroom* (pp. 118–138). Heinemann.

Glowiak, M., & Mayfield, M. A. (2016). Middle childhood: Emotional and social development. In D. Capuzzi & M. D. Stauffer (Eds.), *Human growth and development across the lifespan: Applications for counselors* (pp. 277–306). Wiley.

Goldenberg, E. P., Carter, C., Mark, J., Nikula, J., & Spencer D. (2017). What is repeated reasoning in MP? *Mathematics Teacher, 110*(7), 506–512.

Greater Good Science Center. (2024, March 20). SEL for students: Social awareness and relationship skills. *Greater Good in Education*. https://ggie.berkeley.edu/student-well-being/sel-for-students-social-awareness-and-relationship-skills/

Gulkilik, H. (2016). The role of virtual manipulatives in high school students' understanding of geometric transformations. In P. S. Moyer-Packenham (Ed.), *International perspectives on teaching and learning mathematics with virtual manipulatives içinde* (pp. 213–243). Springer International Publishing.

Hartono, Y. (2020). Mathematical modelling in problem solving. *Journal of Physics: Conference Series, 1480*, Article 012001.

Howse, B. R., Best, L. D., & Stone, R. E. (2003). Children's decision making: The effects of training, reinforcement and memory aids. *Cognitive Development, 18*, 247–268.

Humphreys, C., & Parker, R. (2015). *Making number talks matter: Developing mathematical practices and deepening understanding, grades 4–10*. Stenhouse Publishers.

Illustrative Mathematics. (n.d.). *Anna in D.C.* Retrieved May 13, 2024, from https://tasks.illustrativemathematics.org/content-standards/7/EE/B/3/tasks/997

Illustrative Mathematics. (2024, May 7). *Information gap graphic*. https://hub.illustrativemathematics.org/s/k12/k12-information-gap-graphic

Institute of Education Sciences. (2022). *2022 school pulse panel*. U.S. Department of Education. https://ies.ed.gov/schoolsurvey/spp/

International Center for Academic Integrity. (2021). *The fundamental values of academic integrity* (3rd ed.). https://academicintegrity.org/aws/ICAI/pt/sp/values

Jimerson, S. R., & Haddock, A. D. (2015). Understanding the importance of teachers in facilitating student success: Contemporary science, practice, and policy. *School Psychology Quarterly, 30*(4), 488–493. https://doi.org/10.1037/spq0000134

Jones, S. M., Brush, K. E., Ramirez, T., Mao, Z. T., Marenus, M., Wettje, S., Finney, K., Raisch, N., Podoloff, N., Kahn, J., Barnes, S., Stickle, L., Brion-Meisels, G., McIntyre, J., Cuartas, J., & Bailey, R. (2021). *Navigating social and emotional learning from the inside out*. Wallace Foundation.

Joswick, C., & Taylor, C. N. (2022). Supporting SEL competencies with number talks. *Mathematics Teacher: Learning & Teaching PK-12, 115*(11), 781–792. https://doi.org/10.5951/mtlt.2021.0347

Kagan, S., & Kagan, M. (2009). *Cooperative learning*. Kagan Publishing.

Kaiser, G., & Stender, P. (2013). Complex modelling problem in cooperative learning environments self-directed learning environments. In G. Stillman, G. Kaiser, W. Blum, & J. Blum (Eds.), *Teaching mathematical modelling: Connecting to research practice* (pp. 277–294). Springer.

Kansas State Department of Education. (2017). *Kansas math standards grades K–12*. Author.

Kaplinsky, R. (2016, August 1). *Robert Kaplinsky real world lessons*. https://robertkaplinsky.com/lessons/

Kelemanik, G., Lucenta, A., & Creighton, S. (2016). *Routines for reasoning: Fostering the mathematical practices in all students*. Heinemann.

Koestler, C., Felton-Koestler, M. D., Bieda, K., & Otten, S. (2013). *Connecting the NCTM process standards and the CCSSM practices*. The National Council of Teachers of Mathematics.

Knudsen, J., Lara-Meloy, T., Stevens, H. S., & Rutstein, D. W. (2014). Advice for mathematical argumentation. *Mathematics Teaching in the Middle School, 19*(8), 494–500. https://www.jstor.org/stable/10.5951/mathteacmiddscho.19.8.0494

Knuth, E. J., Stephens, A. C., McNeil, N. M., & Alibali, M. W. (2006). Does understanding the equal sign matter? Evidence from solving equations. *Journal for Research in Mathematics Education, 37*(4), 297–312. http://www.jstor.org/stable/30034852

Kohn, A. (1993). The impact of mathematical modeling on classroom practices. *Journal of Mathematical Education, 14*(2), 125–136. https://doi.org/10.1234/jme.1993.01425

Lane, J. J. (2018). Swinging the pendulum towards social emotional support (a position paper). *The Advocate, 23*(5), 5. https://doi.org/10.4148/2637-4552.1010

Lane, J. J., McKeeman, L., & Bonella, L. (2020). Helping the helpers: Tending to Kansas educators' social-emotional needs and self-care during a pandemic. *The Advocate, 26*(1), 3. https://doi.org/10.4148/2637-4552.1152

Langreo, L. (2023, July 7). *What new research shows about the link between achievement and SEL*. https://www.edweek.org/leadership/what-new-research-shows-about-the-link-between-achievement-and-sel/2023/07

Larbi, E., & Mavis, O. (2016). The use of manipulatives in mathematics education. *Journal of Education and Practice, 7*(36), 53–61.

Lee, H., & Herner-Patnode, L. M. (2007). Teaching mathematics vocabulary to diverse groups. *Intervention in School & Clinic, 43*(2), 121–126.

Lesh, R., & Harel, G. (2003). Problem solving, modelling and local conceptual development. *Mathematical Thinking and Learning: An International Journal, 5*(2/3), 157–190.

Lesh, R., Post, T., & Behr, M. (1987). Representations and translations among representations in mathematics learning and problem solving. In C. Janvier (Ed.), *Problems of representations in the teaching and learning of mathematics* (pp. 33–40). Lawrence Erlbaum Associates.

Life Skills Group. (2018, October 31). *5 ways to improve young children's decision-making*. Life Skills Group.png. https://www.lifeskillsgroup.com.au/blog/5-ways-to-improve-young-childrens-decision-making

Luo, L., & Yu, J. (2015). Follow the heart or the head? The interactive influence model of emotion and cognition. *Frontiers in Psychology, 6*, Article 573. https://doi.org/10.3389/fpsyg.2015.00573

Marshall, S. (2017). A sense of possibility: Cultivating perseverance in an urban mathematics classroom. *Journal of Teacher Action Research, 3*(3), 1–23.

Matney, G., Miranda, F., Knapke, S., & Murray, M. (2020, March 5–7). *Lesson study and teacher's dialog about SMP 5* (Conference session). Proceedings of the 47th Annual Meeting of the Research Council on Mathematics Learning, Las Vegas, NV (pp. 100–107). Bowling Green State University.

Meyer, D. (2013a, May 8). *Teaching with three-act tasks: Act one*. https://blog.mrmeyer.com/2013/teaching-with-three-act-tasks-act-one/

Meyer, D. (2013b, May 13). *Teaching with three-act tasks: Act two*.

REFERENCES

https://blog.mrmeyer.com/2013/teaching-with-three-act-tasks-act-two/

Meyer, D. (2013c, May 14). *Teaching with three-act tasks: Act three & sequel.* https://blog.mrmeyer.com/2013/teaching-with-three-act-tasks-act-three-sequel/

Miller, G. (2023, October 30). *Helping kids make decisions.* Child Mind Institute. https://childmind.org/article/helping-kids-make-decisions/

Monroe, E. E., & Orme, M. P. (2002). Developing mathematical vocabulary. *Preventing School Failure, 46*(3), 139–142.

National Center for Education Statistics. (2022, July 6). *More than 80 percent of U.S. public schools report pandemic has negatively impacted student behavior and socio-emotional development* [Press release]. https://nces.ed.gov/whatsnew/press_releases/07_06_2022.asp#:~:text=Specifically%2C%20respondents%20attributed%20increased%20incidents,COVID%2D19%20pandemic%20and%20its

National Council of Teachers of Mathematics. (2000). *Principles and standards for school mathematics.* Author.

National Council of Teachers of Mathematics. (2014). *Principles to actions: Ensuring mathematical success for all.* Author.

National Council of Teachers of Mathematics. (2017). *Taking action: Implementing effective mathematics teaching practices.* Author.

National Research Council. (2001). *Adding it up: Helping children learn mathematics.* The National Academies Press. https://doi.org/10.17226/9822

O'Connell, S., & SanGiovanni, J. (2013). *Putting the practices into action: Implementing the common core standards for mathematical practice K-8.* Heinemann.

Oles, P. K., Brinthaupt, T. M., Dier, R., & Polak, D. (2020). Types of inner dialogues and functions of self-talk: Comparisons and implications. *Frontiers in Psychology, 11*, Article 227. https://doi.org/10.3389/fpsyg.2020.00227

Pew Research Center. (2024). *Social media fact sheet.* https://www.pewresearch.org/internet/fact-sheet/social-media/

Pierce, M. E., & Fontaine, L. M. (2009). Designing vocabulary instruction in mathematics. *The Reading Teacher, 63*(3), 239–243. https://doi.org/10.1598/RT.63.3.7

Prediger, S., Erath, K., & Moser Opitz, E. (2019). The language dimension of mathematical difficulties. In A. Fritz, V. Haase, & P. Räsänen (Eds.), *International handbook of mathematical learning difficulties. From the laboratory to the classroom* (pp. 437–455). Springer.

Price-Mitchell, M. (2015, June 9). *Creating a culture of integrity in the classroom.* Edutopia. https://www.edutopia.org/blog/8-pathways-creating-culture-integrity-marilyn-price-mitchell

Prinstein, M., & Ethier, K. A. (2022). Science shows how to protect kids' mental health, but it's being ignored. *Scientific American Magazine, 327*(3). https://doi.org/10.1038/scientificamericanmind0922-25

Puentedura, R. (2014). *Building transformation: An introduction to the SAMR model* [Blog post]. http://www.hippasus.com/rrpweblog/archives/2014/08/22/BuildingTransformation_AnIntroductionToSAMR.pdf

Riegel, K. (2021). Frustration in mathematical problem-solving: A systematic review of research. *STEM Education, 1*(3), 157–169. https://doi.org/10.3934/steme.2021012

Rimes, B. (2012, March 1). *Video story problems 101.* The Tech Savvy Educator. https://www.techsavvyed.net/archives/2352

Romrell, D., Kidder, L. C., & Wood, E. (2014). The SAMR model as a framework for evaluating mLearning. *Journal of*

Asynchronous Learning Networks, 18(2), 79. https://doi.org/10.24059/olj.v18i2.435

Rumack, A. M., & Huinker, D. (2019). Capturing mathematical curiosity with notice and wonder. *Mathematics Teaching in the Middle School, 24*(7), 394–399. https://doi.org/10.5951/mathteacmiddscho.24.7.0394

Rumsey, C., & Langrall, C. W. (2016). Promoting mathematical argumentation. *Teaching Children Mathematics, 22*(7), 412–419. https://www.jstor.org/stable/10.5951/teacchilmath.22.7.0412

SanGiovanni, J. (2015, March 23). *Use appropriate tools strategically*. Heinemann, The Standards for Mathematical Practice. https://blog.heinemann.com/smp5

SanGiovanni, J. J. (2017). *Mine the gap for mathematical understanding: Common holes and misconceptions and what to do about them*. Corwin.

SanGiovanni, J. J., Bay-Williams, J. M., Martinie, S., & Suh, J. (2021). *Figuring out fluency - Addition and subtraction with fractions and decimals*. Corwin.

Saucedo, M. (2024, March 28). *Teach your students to look for and express regularity in repeated reasoning*. Elevated Achievement Group, Inc. https://elevatedachievement.com/articles/teach-your-students-to-look-for-and-express-regularity-in-repeated-reasoning/

Schleppegrell, M. J. (2007). The linguistic challenges of mathematics teaching and learning: A research review. *Reading & Writing Quarterly, 23*(2), 139–159. https://doi.org/10.1080/10573560601158461

Schneider, W., & Artelt, C. (2010). Metacognition and mathematics education. *ZDM Mathematics Education, 42*, 149–161. https://doi.org/10.1007/s11858-010-0240-2

Science Buddies. (2024, February 20). *Surface science: Where does a basketball bounce best?* Scientific American. https://www.scientificamerican.com/article/surface-science-where-does-a-basketball-bounce-best/

Scieszka, J., & Smith, L. (1995). *Math curse*. Viking.

Shah, P., Weeks, H., Richards, B., & Kaciroti, N. (2018). Early childhood curiosity and kindergarten reading and math academic achievement. *Pediatric Research, 84*, 380–386.

Sriraman, B., & Umland, K. (2020). Argumentation in mathematics education. In S. Lerman, (Ed.), *Encyclopedia of mathematics education* (pp. 63–66). Springer. https://doi.org/10.1007/978-3-030-15789-0_11

Swanson, E., Ash, J., Cummings, K., Kane, T. J., Staiger, D. O., & Sanbonmatsu, L. (2024). *Come back . . . be here: Evaluating strategies to improve student attendance through a rural research network*. Center for Education Policy Research, Harvard University & National Center for Rural Education Networks. http://ncrern.provingground.cepr.harvard.edu

The University of Texas at Austin: Charles A Dana Center. (n.d.). *Social and emotional learning and mathematics*. Inside Mathematics. https://www.insidemathematics.org/common-core-resources/mathematical-practice-standards/social-and-emotional-mathematics-learning

Tod, D. A., Hardy, J., & Oliver, E. (2011). Effects of self-talk: A systematic review. *Journal of Sport and Exercise Psychology, 33*(5), 666–687. https://doi.org/10.1123/jsep.33.5.666

Tomlinson, C., & Imbeau, M. B. (2023). *Leading and managing a differentiated classroom* (2nd ed.). ASCD.

Tuva Support. (2021, January 25). *Instructional strategy: Claim-evidence-reasoning [new]*. https://support.tuvalabs.com/hc/en-us/articles/360062339033-Instructional-Strategy-Claim-Evidence-Reasoning-new#:~:text=Claim%2DEvidence%2DReasoning%20(CER,it%20supports%20a%20provided%20claim

REFERENCES

Van de Walle, J. (2007). *Elementary and middle school mathematics: Teaching developmentally*. Pearson Education Inc.

Vanoli, L., & Luebeck, J. (2021). Examining errors and framing feedback. *Mathematics Teacher: Learning and Teaching PK-12, 114*(8), 616–623. https://doi.org/10.5951/mtlt.2020.0356

Vocabulary.com. (n.d.). Structure. *Vocabulary.com Dictionary*. Retrieved May 23, 2024, from https://www.vocabulary.com/dictionary/structure

Willingham, J. C., Strayer, J. F., Barlow, A. T., & Lischka, A. E. (2018). Examining mistakes to shift student thinking. *Mathematics Teaching in the Middle School, 23*(6), 324–332. https://doi.org/10.5951/mathteacmiddscho.23.6.0324

Yackel, E., & Cobb, P. (1996). Sociomathematical norms, argumentation, and autonomy in mathematics. *Journal for Research in Mathematics Education, 27*(4), 458–477. https://doi.org/10.5951/jresematheduc.27.4.0458

Zarowski, B., Giokaris, D., & Green, O. (2024). Effects of the COVID-19 pandemic on university students' mental health: A literature review. *Cureus, 16*(2), Article e54032. https://doi.org/10.7759/cureus.54032

Zwiers, J., Dieckmann, J., Rutherford-Quach, S., Daro, V., Skarin, R., Weiss, S., & Malamut, J. (2017). *Principles for the design of mathematics curricula: Promoting language and content development*. Stanford University, UL/SCALE. http://ell.stanford.edu/content/mathematics-resources-additional-resources

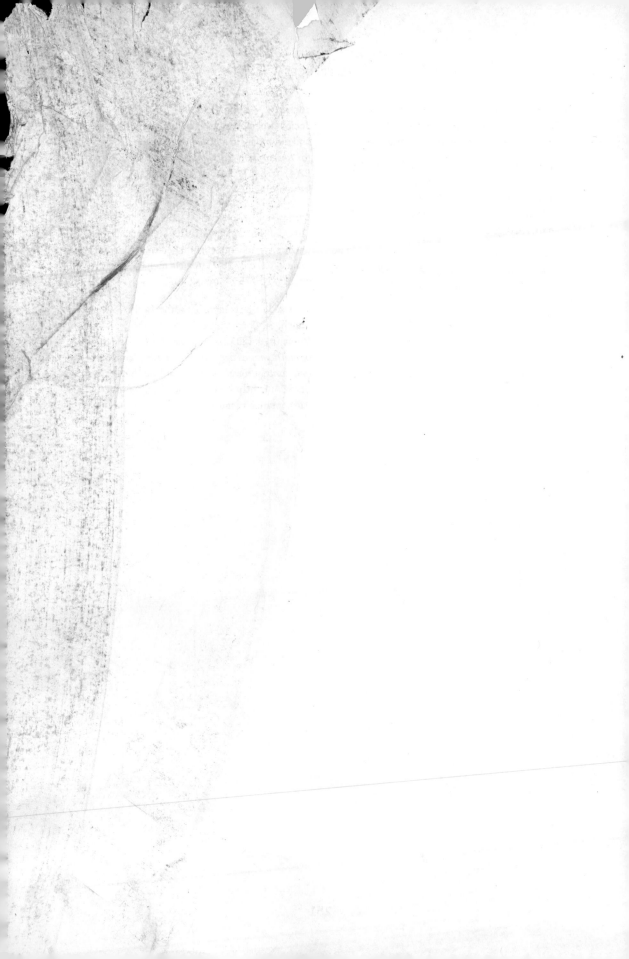

INDEX

2-by-2 tables, 106

abilities to connect concepts to procedures, 213
ability, 8–9, 24–27, 32–35, 45, 53–55, 67, 76, 84–85, 161–62, 166, 174–76, 188, 217–18, 220, 240–41
abstract reasoning, 45, 65
academics, 4–5
accuracy, 149, 161, 165, 167, 169–70, 175
actions, 18, 22, 69, 76–77, 79–80, 85–86, 88–89, 102–3, 112–13, 162, 175–76, 178, 186, 190–91, 203, 236–37, 241–43
Actions Associated, 22, 81, 163, 186
actions students, 88, 134
actions support, 146
active listening skills, 126
activities, next, 208
Act Math Tasks, 31, 121
Adams's mathematics lesson, 99
adaptability, 14, 16, 53, 61–62, 67–68, 108, 117–18, 121–22, 125, 133–59, 191–92, 194, 197–203, 210–12, 222
  explaining, 198
  scaffolds, 206
adaptability and self-regulation, 192, 210–11
adaptability in grouping information, 194
adaptability skills, 59
adding, 24–25, 43, 48, 76, 109, 161, 170, 186, 215, 227–28, 231
addition and subtraction, 24–25, 38–39, 48–49, 72, 215
addition of intrapersonal and interpersonal skills, 16
addition table, 216
address intrapersonal, 130
address intrapersonal and interpersonal skills, 130
Adequate, 37, 68, 95, 202, 226
Adequate Competencies, 96, 124, 150, 177, 226–27
Adequate evidence, 36, 67, 95, 123, 150, 177, 201

Adequate evidence Download, 226
Adequate Math Practice, 123
Adequate Practice, 150, 177
age group, 15, 24, 81, 172
algebraic manipulations students, 50
Algebra Tiles, 87
algorithms, 188, 213–14, 216, 228, 237
Aligning mathematics lessons, 76, 183
answer questions, 158
answers, 22–23, 48–49, 59–60, 72, 74, 118–19, 156–57, 164, 173, 180, 204, 206, 209, 217–18, 228–30
  correct, 18, 28, 155, 167, 209–10
  numerical, 161–63
approaches, 9–11, 22–23, 25–28, 30–31, 42–43, 57–59, 64, 66, 81, 83, 180–81, 188–89, 196–97, 199, 219
  new, 8, 20, 112
  scaffolded, 65, 167
approach modeling, 112
appropriate, 149–50
A.REI, 26
argumentation, 79, 91, 93, 102
arguments, 9, 11, 13, 27, 79–81, 83–87, 89–90, 92, 95, 99, 102–3
  mathematical, 90, 103, 172
arrows, 18
Ask Questions and Formulate Problems, 116
assertiveness, 11, 14, 83–85, 87–88, 94, 96–97, 99, 101–2
Assessing Interpersonal and Intrapersonal Skills, 35, 67, 93, 122, 148, 175, 201, 225
assessment, 2, 14, 17–19, 149, 158, 176, 203
assessment criteria, 18, 175
assessment strategies, 68, 225
assumptions, 26, 80, 112, 121, 128, 130, 134, 138
  comfortable making, 106
attention, 9, 11, 53–54, 57, 72, 74, 166–67, 170, 173, 175, 196, 198, 210–11
Attention to mathematical structure, 211

attributes, 36, 115, 155, 164, 179, 198
attributes in table, 36
authentic situations, 107, 113
awareness, 11, 18, 165, 176–77, 239, 241, 243

behaviors, 3, 6–9, 53, 59, 69, 79, 84, 91, 190–93, 197, 240–42
believing, 31–32, 39
benefits, 20, 23, 34, 114, 136, 147–48, 153, 157
benefits and limitations, 141, 147–48, 153, 155, 159
Benefits and Limitations for Tools, 147
benefits and limitations of tools, 141
blocks, 186, 228–29
board, 70, 73, 99, 129, 180, 197
book, 5, 12, 16, 20, 35, 39, 54, 58–59, 89, 92, 121, 220–21, 239–40
book report, 58–59
borrow, 188, 199–200
bottle, 2-liter, 119
bounce height, 127–28
building, 3, 16, 21, 28, 86, 115, 127, 180, 214, 232, 242
building mathematics lessons, 24

calculations, 35, 46, 53, 156, 169–70, 214
    repeated, 213–14, 217
calculators, 129, 133, 137, 139, 147–48, 155, 232
cards, 97, 194–95, 224
    students match, 194
Card Sorts, 194, 196
CASEL, 6, 241
Catching students, 191
category, 11, 111, 164, 178
    conceptual, 111
CCSSM, 25–26, 38, 40, 46–49, 51, 80, 82–83, 97, 106, 108–9, 136–38, 164–65, 187, 189–90, 214–16
CDC, 2–3
CER (Claim-Evidence-Reasoning), 171
CER Data Stories, 171–72
challenges, 2, 5, 21, 24, 26–27, 33, 35, 39, 41, 44, 47, 184, 219–21
chance, 14, 17, 47, 86, 94, 122, 126–27, 221

change, 16, 18, 22–23, 100, 155, 222
    rate of, 137, 207
change in content and lesson development, 18
charge, 73–74
chart, 89, 147–48, 231
cheat, 139–40
cheating, 156–57
checklist, 69, 149, 173
children, 6, 21, 38–39, 69–71, 79, 84, 89, 111–14
children playing, 69–70
Children's Literature, 35, 54, 121, 220
choice of tools, 137
choices, 83, 88–89, 93, 102, 111–12, 118, 137, 141, 143, 146–47, 158
claim, 87, 89–90, 92, 103, 162, 171–72, 221
    making, 88–89
Claim-Evidence-Reasoning (CER), 171
class, 40–41, 87, 89–90, 113, 125–26, 128–30, 142, 151–53, 155, 157, 178–79, 197–98, 205–6, 228, 241–42
classmates, 92, 102–3, 113, 205, 212, 221
class observation tool, 36, 67, 94–95, 122–23, 149, 152–53, 176, 201, 226, 232
class periods, 113
classroom, 3–7, 35, 38, 40, 72, 74, 92–93, 139, 141, 144–46, 191, 218–19, 232, 236–37, 242
classroom examples of teachers, 158
classroom measurement example in Competency Builder, 149
classroom narratives, 43, 75–76, 102, 130, 183
classroom routines, 43, 183
classroom teachers, 4
collaborative learning strategies, 158
collaborative learning structures support engagement, 85
column, 22, 92, 141, 193, 205, 229
com/lessons, 121
comma, 234
common multiplication, 51
communicate, 84, 86, 113, 161–62, 166, 173
communication, 9–11, 14, 36–37, 43, 47, 53, 56, 75, 77, 165–68, 173, 175, 177–78
communication skills, 10, 27, 69, 72, 83, 91, 162, 166, 176, 180, 182–83
    sharpen interpersonal, 10

INDEX

communication skills and empathy, 91
communication skills and teamwork, 10
commutative property, 222
Comparing Tasks to Encourage Appropriate and Strategic Use of Tools, 143
competencies, 6–7, 37–38, 138, 145, 151, 201–3, 227, 232, 237, 239, 241–42
competencies students, 133
   social-emotional, 242
Competency, 8–9, 11, 36–37, 67–68, 85, 95–96, 124, 149–51, 178, 201, 203
Competency Builder, 30–35, 54–55, 57, 59, 85–86, 88–89, 114–15, 117–18, 121–22, 139, 146–47, 168, 198–99, 218–20, 222–24
Competency Mathematics, 95, 149–50, 201, 226
complement, 16
complex problems, 22, 191
   down, 188
computational concept, 139
conceptual understanding of equations, 51
Concrete models support, 86
conflicts, 79, 86
conjecture, 91–93
connections, 3, 5, 8, 10, 12, 17, 52, 64–65, 136, 138, 165–66, 190–91, 195, 210–11, 217
connections to social-emotional learning, 30
consistencies, 5, 197, 236
constructing, 79, 84, 86, 107, 131
constructing arguments, 80, 82, 86, 89, 94, 103, 174
Constructing Claim-Evidence-Reasoning, 171
construct viable arguments and critique, 84–85, 102
content, 13, 15–16, 18, 76, 83, 88, 93, 130, 155, 158–59, 163, 216–17
content standards, 12–13, 18, 25, 29, 83, 102, 106, 108–9, 111, 137, 187, 190, 215–16
content standards and mathematical modeling, 111
context, 22, 27–28, 46, 72–73, 76, 106–8, 110–13, 117, 120, 122–23, 133, 136, 138
   problem's, 49, 64

context of everyday problem-solving scenarios, 22
contextualize, 45–47, 55, 67, 76
conversations, 39–40, 72, 77, 101, 126, 209–10
coordinate plane, 83, 136, 155, 187, 190, 215, 232
coordinates, 136–37, 155, 234
Corners Organizer, 168–69
correspondences, 22, 162, 172
Corresponding Mathematics Goal, 24, 48, 82, 108, 136, 163, 189, 215
costs, 23, 52, 73, 75, 126, 180–82
count, 135, 228–29
counterexamples, 80, 91–93
   producing, 91
counters, 38–39, 135, 222, 228
   red, 38
   yellow, 38
counting, 35, 63, 72
courage, 139–40, 146
crate, 35
creative thinking, 108, 112, 115, 117–18, 121, 125, 127, 129–30, 214–15, 217–18, 222–23, 225, 237
creative thinking and adaptability, 117–18, 222
Creative thinking Launch/Introduction, 13
creative thinking skills, 115, 118
creativity, 122, 124, 128, 226–28, 232, 236
critical thinking, 28, 52, 118, 219, 225
critical thinking skills, 82, 219
critique, 13, 79–81, 84–85, 87, 95, 102, 121
critiquing arguments, 79, 91
critiquing mathematical arguments, 79
cubes, 119, 135, 227
cups, 61, 63
curiosity, 118, 120, 128–29, 180–83, 211, 214–15, 217–19, 225, 227, 232, 235–37
curiosity and creative thinking, 129, 214, 225
curiosity-based questions, 218
curiosity in students, 118
Curiosity Interpersonal Skills, 13

data card, 59–61
day, 5–6, 32, 58–59
   first, 58–59
decimals, 164, 170, 180–81, 189, 203, 211
   moving the, 189

decision-maker, good, 154
decision-making, 9, 11, 13, 81, 83–84, 87, 89, 91, 94, 96, 109, 111, 114–15, 124–25, 149–51
decision-making process, 113–14, 130, 138
decision-making process and social awareness, 113
decision-making situations, 148
decision-making skills, 10, 133, 145, 151
decisions, 89, 94, 100, 112, 114–15, 124, 127, 130–31, 141, 145, 155
   making, 88, 106, 114–15, 148–49, 158
   strategic, 137
   strategic mathematical, 9
decomposing, 96–97
decompositions, 82–83, 97–98
decontextualize, 45–46, 53, 55, 76, 123
deeper understanding, 26, 146, 161, 168, 182, 211
defining social awareness, 85
definitions, clear, 161–62
degree of precision, 162–63
demands, 1–2
denominators, 82, 96, 164, 186
descriptions, 7–9, 30, 86, 137–38, 147, 166, 212
development, 8, 10–12, 14–15, 27, 35–36, 88–89, 102, 106, 108, 122, 138, 147–49, 162, 215–18, 236–37
development of confidence and curiosity in mathematics, 183
development of integrity, 139, 141, 148
development of integrity in learners, 139
development of intrapersonal and interpersonal skills, 69, 175
development of social awareness, 85, 91, 130
development of social-emotional competencies and MP8, 220
diagonals, 155
diagrams, 8, 22, 80, 86–87, 106, 112, 120, 133, 147, 161
differences, 1, 50, 92, 117, 129, 157, 170, 198, 208, 210
Differences Strategy, 170–71
digits, 161, 186, 189–90, 206, 231
dimensions, 25, 35, 221
discussing mathematical ideas and strategies, 27

discussing models and solutions, 131
display questions students, 90
distance, 40, 156–57, 235
division, 51–52, 137
division situations, 51
Download, 9, 11, 29–30, 33–34, 36–37, 50, 52, 65–68, 90, 92–93, 95–96, 150–51, 177–78, 201–3, 227
Download Figure, 30

edges, 135, 149, 179
education, 1–3, 5, 15, 147, 175
educational purposes, 89–90
educators, 3–5, 139, 241
EE, 48, 51, 76, 164–65, 180, 183, 187, 190, 216, 232
effectiveness, lesson's, 42, 182
efficiency, 135, 137–38, 141
Eighth-Grade Example, 58
Elementary Example, 199
Elevated Achievement Group, 167
emotions, 8–10, 18, 47, 69, 72, 91, 113, 176, 191–93, 203, 209, 211, 242
emotions and actions, 18, 69, 176
empathy, 11, 14, 81, 83–85, 87, 91, 94, 96–97, 99, 101–2, 106
engagement, 35–37, 67–68, 86, 88, 93–95, 118, 120, 122–23, 134–35, 137, 148–50, 163–64, 175–77, 201–2, 225–26
engagement in MP2, 71
engagement in MP2 and social-emotional competencies, 71
engagement in MP5, 149
engagement in MP5 and development, 149
engagement in MP5 and development of social-emotional competencies, 149
Engagement Strategies, 17, 28, 53, 84, 113, 138, 166, 192, 218
environment, supportive learning, 24, 27, 206
equations, 25–26, 45–46, 48–49, 51, 66, 71–76, 82–83, 88, 101, 109, 137, 156–57, 164–65, 181–82, 186–87, 195–200, 214–16
   linear, 87, 194
equations for real-world situations, 76
equivalence, 195, 205–6
erase, 209–10

INDEX

erasing, 191, 209–10
errors, 8, 11, 81, 116, 153, 168, 171, 218
   possible, 134, 138
estimates, 54–56, 119–20, 147
estimation, 55–56, 134, 138, 164
Estimation Examples, 56
Estimation Exploration, 55–56
evidence, 36–37, 67–68, 84, 89, 94–96, 103, 122–24, 142, 149–50, 171–72, 177, 201–2, 226–27
evidence and support, 84
evidence of learning, 149
evidence of students, 36
evidence to support, 171
example of self-awareness, 99
Example of Student Work Sample, 234
examples, 32, 35–36, 47–49, 51–52, 54–55, 60, 69, 82–83, 89–94, 96–97, 117–19, 124, 148–49, 169–73, 176, 179–80, 185–86, 188–90, 211–13, 222–25
   area, 51
   concrete, 17
   exchange, 222
   generated, 155
   measurement, 52
   pictorial, 61
   real-life, 42
   real-world, 174
Examples for Partial Sums and Differences Strategy, 171
examples highlight, 26, 67
examples of quadrilaterals, 164, 178
examples of shapes, 179
exercise, 35, 110–11
experience, 17, 24, 112, 116, 146, 163, 242
exploration of tools, 159
exponents, 193, 211, 222–23
exposing students, 199
expressions, 25, 38, 66, 165, 185–86, 196, 199, 214, 222–23

factoring, 82–83, 88, 99, 101, 186
fairness, 9, 139–41
familiarity, 220
families, 3, 108, 241–42
feedback, 9, 14, 17–18, 34–35, 67–68, 77, 81, 84, 89–91, 94, 99–102, 142, 148

Figure, 28–30, 55, 59–60, 62–66, 85, 89–91, 107, 110–11, 114–22, 169–73, 194–98, 207, 220–21, 223–24, 229, 234
finding, 21, 26, 39, 42, 54, 129, 167, 196, 217–18, 231–32, 235
finding keywords and numbers, 28
finishes, 58–59
first-grade class, 227
F.LE, 108, 127–28
Flipside, 54
floor plan, 25, 40–42
fluency, procedural, 64, 187
focus, 1–2, 9–10, 15, 25, 27, 52–54, 56–57, 72, 74–76, 83–84, 108, 146–47, 171, 182–83, 189, 193–95, 209
   increasing, 168
focus on place value and structure, 194
focus teacher feedback, 77
formats, make-shift school, 3
formulas, 106, 115, 136, 155, 157, 213–14, 216
   distance, 136, 155–56
   quadratic, 82, 88, 99, 101
Formulate Problems, 116
fostering self-regulation, 47–77
foundation, 16, 49, 106–7, 163
fourth-grade class, 96, 125
fractions, 82, 87, 93, 96–98, 164, 170, 180, 186, 192
   decomposing, 82, 98
   unit, 82, 97
frame mathematical tasks for students, 219
framework, 12, 24, 47, 81–82, 108, 136, 163, 189, 239–40
framework for designing lessons, 189
framework for designing mathematics lessons, 163
friend, 55, 110
front-end planning on social-emotional competencies and enhanced lesson reflection, 18
functions, 49, 53, 105, 108, 111, 129, 134–35, 137, 190, 207–10, 225
   exponential, 108, 127–28
function shapes, 190
function shapes and equations, 190

games, 90, 223–24
game to students, 224
generation skills, next, 1, 239
geometric figures, 25, 40
geometric transformation, 115, 222
goal, 12–13, 22, 25, 33, 36–37, 47–48, 74–75, 82, 108–9, 135–36, 166, 168, 174, 176–77, 215–16
grades, 56, 61, 80, 90, 109, 133, 196
graph equations, 165
Graphic Organizer for Multiple Representations Stemming, 66
graphic organizer in Figure, 120
graphic organizer in small group, 168
graphic organizers, 61, 64–66, 114, 120, 133, 168–69, 195
Graphic Organizer to Focus Attention, 168
graph in Figure, 172
graphing, 83, 99–101, 234
graphing calculator, 22, 133–34, 148
graph paper, 232–33
graph reasoning, 66
graphs, 133–35, 137, 147, 172–73, 186, 189–90, 194–95, 198, 207–9, 216, 234–35
graphs and algebraic equations, 190
graphs and tables, 216
graphs of functions and solutions, 134
grass, 49
Greater Good Science Center, 85
ground, 58, 234
groupings, 98, 194
group interactions, 87
group of children, 69–70
groups, 39, 41, 51–52, 97–98, 100, 103, 121, 125–30, 152–53, 155, 228–29
  first, 70
  small, 64, 90, 97, 122, 125, 128, 142, 168, 199
Group Size Unknown Unknown, 51–52
groups of students, 121
groups share, 122, 229
group students, 87
growth, 32, 34, 42, 94, 122, 158, 182, 203
guide, 1, 39, 44, 100, 141, 173, 184
guided students, 180
guidelines, 145–46
guide student reflection, 176, 225

guide students, 26, 48, 113
guide students in making sense of problems, 48

height, 128, 145, 232–33, 235
Heinen nods, 155–56
helping students, 89, 137, 140, 197, 211, 236
helping students pass, 2
helping students practice, 242
helping students process, 47
High-quality mathematics tasks, 28
High-Quality Task Rating Tool, 30–31
high-quality tasks, 28–31, 158, 183
High School Algebra Class, 99, 207
high school students, 134
hinders students, 213
holistic lessons, developing, 14
hundredths, 194, 206

identifying mathematical patterns and structures, 188
Identifying Problem-Solving Approaches, 114
Illustrative Mathematics, 31, 56, 61, 74, 180, 196
implementation of tools, 146
implementing competency builders, 239
implicit intrapersonal, 16
implicit intrapersonal and interpersonal skills, 16
Improved adaptability and flexibility, 240
improvement, 9, 42, 182
impulsive problem-solving, preventing, 57
inches, 40–42, 51, 144–45, 232
inches of string, 51
income, unstable, 3
Incorporating intrapersonal and interpersonal skills, 211
increasing focus on precision, 168
individual observation tools, 36, 150, 176
individual student observation tool, 37, 68, 94–95, 122–23, 177, 202, 225–26, 228
information, 35, 56–60, 64–65, 80–81, 112, 114, 118–22, 130, 147–48, 157, 192–93, 204, 206, 232, 234
  grouping, 194
  missing, 61, 64
  new, 11, 112, 120, 127, 131, 191, 197–98, 212

INDEX

Information Gap, 59–61
Information Gap Problem Card and Data Card, 61
Information Gap Student Conversation, 60
information students, 119, 193
Input-Output Table, 224
instructing students, 170
Instructional Routines, 196
instructional strategies teachers, 236
Instructional Structures, 17, 28, 53, 84, 113, 138, 166, 192, 218
instructional tools and methods, 43, 131
instructional tools and methods to make, 131
instruction tools and methods, 75, 102
integration and value of intrapersonal and interpersonal skills, 43
integration of intrapersonal and interpersonal skills, 76, 138, 183
integrity, 9, 13, 135, 138–39, 141–42, 147–52, 155–59
   promoted student, 154
   tools Responsible, 152
interpersonal, 7, 10, 19, 35, 79, 82, 175, 178, 225, 235–36, 241
interpersonal competencies, 1, 67, 212
interpersonal learning competencies, 43, 75, 102
interpersonal skill development, 75, 77, 154
interpersonal skills, 6–7, 10–14, 16–17, 26–27, 42–43, 47, 52–53, 76–77, 80–81, 83–84, 111–12, 135, 138, 157–59, 165–66, 182–84, 191, 215, 217, 236–37
   embed, 217
   identifying, 16
interpersonal skills of creativity, 236
interpersonal skills of goal setting, 236
interpersonal skills students, 162
interpersonal skills to highlight, 10
Interpersonal Standard, 36, 176–77
intrapersonal, 1, 7, 9–14, 18–20, 43, 75, 77, 79, 81, 176–77, 181, 183–84, 211–12
intrapersonal and interpersonal connections, 12, 20, 200
intrapersonal and interpersonal connections by affirming student actions, 200
intrapersonal and interpersonal learning competencies, 43, 75, 102

intrapersonal and interpersonal skills, 6–7, 12, 16–17, 26, 42–43, 52, 69, 76–77, 80–81, 83, 111–12, 135, 157–59, 165, 175–76, 182–83, 201–2, 217
intrapersonal and interpersonal skills in math, 157
intrapersonal and interpersonal skills in nonmathematical situations, 77
intrapersonal and interpersonal skills students, 162
intrapersonal and interpersonal skills to support, 77
intrapersonal communication, 47, 53, 72, 75–76
intrapersonal competencies, 94, 122
Intrapersonal Interpersonal, 95, 149–50, 201
Intrapersonal/ Interpersonal Skills, 14
intrapersonal skills, 8, 10, 13, 16, 27, 35–36, 82, 84, 111–12, 137–38, 175, 191, 215, 217, 235–36
   developing, 8
   selected, 165
intrapersonal skills of curiosity, 236
inverses, 54
inverse variation, 208
involving addition, 26, 164
involving addition and subtraction of fractions, 164
involving situations, 25, 48, 109
involving students, 188
involving students in MP7, 188
items, 143, 145, 217
Izzie, 208–10

41
judgments, 103, 106, 112
justification, good, 89

Kenyon regroups students, 70
key concepts of intrapersonal and interpersonal skills, 7
key strategies for problem-solving in mathematics, 21
kids, 71
Kindergarten Example, 58

labels, 110, 144, 165, 167–68, 173–74, 229
language, 46, 48, 83–85, 103, 109, 136–37, 153, 161, 164, 190, 216

language frames, 71–72, 77, 91, 103
layer, critical, 6
learners, 1, 14–15, 17, 20, 28, 42, 139–40, 182, 192, 214, 218
learning, 3–6, 32–34, 42, 62, 90, 102, 117–18, 149, 156–57, 169–71, 182, 184–85, 217–18, 224–25, 235–36, 240–43
   cognitive, 14, 18
   peer-supported, 206
learning and flexibility in problem-solving strategies, 62
learning and problem-solving, 171
learning experience, 42, 130, 139, 182
learning goals, 99, 201
learning math, 16, 32, 240, 242
learning process, 8, 43, 139, 143, 165, 171
length, 40–41, 51, 136, 143–45, 151, 155, 179, 232
   actual, 25, 40–42
lesson, 5–7, 10, 12–18, 20, 24–27, 42–44, 91, 93–94, 101–3, 108, 136, 139–40, 154–55, 162–63, 174–76, 182–84, 189, 215–16, 225, 239–40
   building, 102
   classroom, 43, 76, 130, 182, 189
   designed, 102
   designing, 189, 210
   developing, 18
   favorite, 20
   first, 232
   fraction, 87
   hard, 55
   holistic, 19–20
   identifying, 174
   individual, 225
   modeling, 108, 117
   next, 30
   plan, 80
   seventh-grade, 40
   upcoming, 77, 211–12
lesson activities, 14, 240
lesson content, 191, 217
lesson development, 18, 24, 163
lesson goal, 48, 108
Lesson Learned, 33
lesson plan development, enhanced, 12

lesson planning, 10, 12, 14–15, 20, 43, 189, 192
   careful integrative, 239
   integrative, 6
lesson planning process, 15, 18, 183
lesson plans, 1, 18, 46, 77, 141, 212
   strong, 6
   traditional, 16
lesson reflection, enhanced, 18
lesson's content goals, 240
lesson's content standards and goals, 47
leverage, 81–82, 86, 88, 121, 133, 146, 162, 166, 190, 225, 242
leverage MP4, 108, 130
leverage number sense, 196
leveraging adaptability, 185–211
Life Skills Group, 114
limitations, 114, 133, 136, 141, 147–48, 153–55, 157, 159
limited decision time, 113
list, 85, 103, 114, 120, 124–25, 128–29, 196, 213, 224, 231, 234
log, 33–34

Maier's algebra students, 211
Maier's students, 208
make-shift school formats and structures, 3
make use of structure, 185–86, 192, 197, 199, 210, 219
making sense of problems, 21, 26–27, 35–36, 43, 48
making sense of problems and persevering in solving, 21, 26, 35–36, 43
manipulatives, 137, 146
mantra, 54–55
matches Card, 195
math, 1–2, 10, 13, 16–19, 34, 107, 110–11, 114, 122, 157, 242
math class, 108, 114, 127, 213
math classroom, 10, 12, 113, 242
math content, 15–16, 18–20
math content knowledge and engagement in MP4, 122
math content standards, 14
math curse, 121
mathematical, 13, 37, 46, 79, 82, 107, 186
mathematical abilities, student's, 166

INDEX

mathematical communication, precise, 161, 183
Mathematical Content Standard, 24, 48, 82, 108, 136, 163, 189, 215, 237
mathematical evidence, 89
mathematical goals, 13–15, 25, 136, 166
mathematical ideas and strategies, 10
mathematical modeling, 105–7, 109–11, 118, 121, 127
mathematical modeling problems, 111
mathematical models, 11, 107, 111, 120–22, 130–31, 138, 174
mathematical objects, 198
mathematical patterns, identifying, 188
mathematical practices (MP), 8, 12–21, 24–25, 27–28, 35–36, 46–48, 81, 83, 108–9, 135, 137, 163–64, 183, 189–91, 215–16
mathematical practices of making sense of problems and persevering, 35
mathematical practice standards, 38, 69, 96, 125, 151, 178, 203, 227
mathematical problem for understanding, 56
mathematical problem in table, 110
mathematical problems, 8–9, 48, 50, 107, 109–10, 112, 131, 133, 138, 164–65, 188, 215, 220
  appropriate, 137
  down, 64
  real-world, 183
mathematical problem-solving, 9, 22–23, 27, 53
mathematical questions, 57–59, 70, 73, 117–18, 120
mathematical questions and reason, 120
mathematical representation and number, 65
mathematical routines, 192, 197
mathematical situations, 9, 106, 147, 215
  real-world, 45
mathematical standards, selecting, 189
mathematical tasks, 28, 69, 111, 159, 237
  frame, 219
  high-quality, 117, 236
mathematical tasks students, 27, 165, 191, 217
mathematical thinking and reasoning, 51, 88
mathematical tools, 133–59
  appropriate, 137

mathematical understanding, 88, 183
mathematical understanding and communication skills, 183
mathematical vocabulary, 161, 175
mathematicians, 161, 204, 213–14
mathematics, 15, 21–24, 37–38, 43, 45–48, 64–65, 67–68, 75–77, 79, 89, 102, 105–10, 146, 161–64, 166–67, 176–78, 185–88, 192–94, 209–15, 219–22
  modeling, 106–7, 112
mathematics class, 79, 139–40, 171, 174, 213
mathematics classroom, 32, 44, 47, 54, 83, 133–34, 139, 184, 212, 215
mathematics competencies and understanding of intrapersonal and interpersonal skills, 42, 182
mathematics concepts, 29, 82, 86, 146, 183
mathematics goal, 24–25, 28, 82, 84, 93, 136, 138, 148–49, 163–64, 189–90, 192, 215, 218, 225
mathematics learning, 189, 242
mathematics lessons, 12, 42–43, 76–77, 89, 103, 108, 162, 183, 202, 211, 217
  designing, 163
  preparing, 108, 135, 163
mathematics practices, 24, 43
mathematics problems, 47, 108, 125, 155, 166, 180
  real-world, 53, 76
  single, 54
mathematics situations, 52
mathematics skills, 34, 44, 67, 184
mathematics tools, 142, 155
math goal, 13, 16–17, 108, 113
math lessons, 12, 16, 108, 130, 239
math practices, 10, 13, 15, 17, 109, 239–40
math practices teaching, 242
math problem, 9, 110–11, 121
  complex, 240
math student, 9, 11, 29–30, 33–34, 36–37, 50, 52, 65–68, 90, 92–93, 95–96, 150–51, 177–78, 201–3, 226–27
Math Talk Examples, 199
Math Talks, 199
math tools, 242
meals, 180–81
meaning, 22–23, 25, 46, 74–75, 107–8, 110, 154–55, 162, 164, 174

measure
  object's, 25
  units of, 144–45, 162, 167–68, 172, 174
measurements, 41, 51–52, 115, 128, 144–45, 149, 152–54
measure time intervals, 26
meat, 117
mental health, 2–3, 241
mental wellness, 4, 6
merge, 239
merges content standards, 12, 239
merging content standards, 24, 47, 82, 108, 135, 163, 189, 215
meterstick, 152, 154
methods, 14, 17, 22–23, 28, 82–84, 88, 99–102, 113–14, 128, 131, 170, 214–15, 218
  general, 213–14
methods for solving tasks, 28
methods in small groups, 128
methods students, 99
middle school students, 214
mindset, 21, 23, 27, 33–34, 47, 54, 82
minutes, 26, 88, 100–101, 113, 128–29, 152, 232
mistakes, 32–34, 72, 87, 103, 154, 169, 174, 240
model curiosity, 218, 231
modeler to ask questions, 106
modeling, 24, 79, 105–7, 110–12, 115, 122, 125, 130, 167, 220, 242
modeling activities, 113, 121
modeling cycle, 107, 110, 121–22
modeling practices, 130
modeling problems, 110–11, 130
modeling process, 9, 105, 107–8, 112–13, 116, 118, 122, 127, 131
modeling situations, 109, 112, 116
  real-world, 111
modeling skills, 121
  developing mathematical, 107
models, 6, 48–49, 72, 80, 82–83, 86–87, 103, 105–8, 112–13, 115, 117, 120–21, 131, 137, 228–29
  best, 108, 129
  concrete, 87, 115, 133, 215
  discussing, 131
  fraction, 97

  linear, 195
  real-world, 107, 111, 130
model situations, 120, 130
money, 6, 61, 73–75
monitor, 9, 22, 25, 57, 101, 122, 169, 173, 188, 191, 194
monitor skill development, 68, 202
monitor students, 159, 237
Montgomery, 110
MP. *See* mathematical practices
MP1, 21–25, 27–28, 36, 38, 43–44
MP1 Problem, 13
MP2, 46–48, 52–54, 61, 69, 71–72, 76, 107, 109
MP3, 79, 81, 83–86, 92, 94, 97, 99, 102
MP4, 106–7, 109, 111, 113, 122, 125, 130
MP5, 107, 133–35, 137–39, 141, 143, 149, 151, 158–59
MP5 and intrapersonal and interpersonal skills, 135
MP5 states students, 147
MP5 Student Actions, 134
MP6, 161–66, 178, 181–84
MP7, 186–92, 194, 196, 198, 210–11, 214
MP7 and intrapersonal and interpersonal skills, 189
MP7 and social-emotional competencies, 211
MP7 by exposing students, 199
MP7 by exposing students to multiple problem-solving strategies, 199
MP8, 213–21, 224–25, 235–37
multiple angles and adaptability in grouping information, 194
multiple representations and graphic organizers in support, 61
Multiple Representations Poster, 62–63
Multiple Representations Stemming, 66
multiple strategies and tools, 23
multiplication table, 216
multiplying, 73, 189, 203, 206
multi-step, 164, 180–81

Name Goal, 67, 95, 123, 149–50, 201, 226
naming, 16, 18, 20, 24, 77, 145, 191, 197
natural connections of MP2 to intrapersonal and interpersonal skills, 47
NCES, 3–4
NCLB (No Child Left Behind), 2

INDEX

NCTM, 46–47, 69–71, 146, 162, 187, 213
newspaper, 92
New York Times, 173
next time, 35, 202, 209
   improved, 174
No Child Left Behind (NCLB), 2
nonexamples, 155
nonmathematical situations, 21, 77
non-traditional units of measurement, 154
notecard, 99–100
number answers, 48
number bonds, 194–95, 228
number exponents, 189, 203
numbers, 25–26, 45–46, 48–49, 51–52, 54–55, 57–59, 70–73, 75–76, 82–83, 93, 161, 164–65, 179–80, 187–90, 192–94, 203–6, 216, 222, 224, 230–32
   adding two-digit, 215, 228
   stair, 234
   subtracting multidigit, 170
   two-digit, 215, 228, 231
   unknown, 26, 48, 109, 137, 165
number sense, 55, 75, 196
number sentence, 46, 65–66, 72, 76, 194, 196, 228–29
numbers mathematical structures in problem, 46
numeral, 193–94

OA, 26, 48–49, 76, 109, 137, 164, 187, 216
objects, 24–26, 46, 48–49, 80, 83, 86, 92, 109, 112, 136, 143, 185–86, 198
online, 110
Online Platform Usage, 172
operations, 25, 46, 48, 74–75, 181, 199, 219
options of tools, 152
order thinking skills, 182
organizer, 147–48, 169
outcomes, 2, 6, 9, 107, 109, 115, 124
   maximize student, 4

packs, 228–29, 231
pages, 35, 58–59
pairs, 39–40, 42, 70, 73, 83, 87, 103, 142, 172, 197, 222
partial sums, 48, 170
Partial Sums and Differences Strategy, 170–71

partner, 32, 35, 97, 101, 139, 142, 144, 195, 198–99, 207, 209, 230, 232–33
path, 32
patterns, 8–9, 45–46, 185, 188–90, 192–93, 200–201, 203–6, 210–12, 214–23, 225, 228, 230, 235–37
   discern, 192, 210
patterns and structure of mathematics, 212
patterns and structures, 188, 200, 211, 214
patterns and structures in mathematics, 214
pattern/structure, 202
pause, 45, 75, 88, 204, 206–7, 209
Peer Examples and Counterexamples, 93
peers, 84, 87, 90–91, 94, 97, 99, 101–2, 141, 145, 167, 171, 174, 217
peers for evidence of social awareness, 94
pencil, 54, 86, 133, 135, 140, 145, 191, 209, 232
perimeter, 93, 108–9, 136
perseverance, 9, 13, 16, 24, 27–29, 31, 33, 36–43, 214–15, 217–21, 225–27, 231–32, 236–37
perseverance and teamwork, 220, 225
perseverance in problem-solving, 28
persevere, 13, 22, 27–29, 34, 38–42, 184, 219–21, 236, 242
persevering, 21, 26, 35–36, 43, 219, 235
persistence, 31–32, 54
perspectives, 11, 27, 30, 56, 84–85, 87, 89, 113, 172–73, 193–94, 198–200, 202, 205, 210–12, 239–40
   diverse, 11, 24, 27, 173, 197–98
phones, 89–90, 232
physical models, 87, 115
pictures, 4, 22, 55, 106, 117, 125, 155, 159, 195, 228, 232
place, 113, 140, 143, 189, 197, 230–31
place value, 190, 194, 206, 211, 215, 227
place value chart, 189, 229
plan, 22, 24, 26, 28, 32, 71, 73–75, 102, 105, 108, 140–41, 162–63, 167
planning, 12, 15, 17, 81, 84, 108, 135, 138–40, 163, 166, 215–16, 218, 239–40
planning mathematics lessons, 24
planning phase of selecting tasks for students, 215
planning process, 24, 141, 240
Planning Questions, 141

player, 224
playing, 69–71, 146
Pointing, 98, 231
positive number times, 185
posters, 62, 64, 162
Power Examples, 223
powers, 189, 193, 203, 206, 211, 217, 223
practice, 15, 17–18, 36–37, 42–44, 67–69, 76–77, 79–80, 82–84, 88–90, 95, 99, 102, 151–53, 162–63, 167, 175–77, 182–84, 201–2, 226, 242
practice adaptability, 212
practice empathy, 103
Practice Engagement, 149–50
Practice Intrapersonal Competency Mathematics, 36
Practice Intrapersonal Interpersonal, 123
practice of mathematical modeling, 105
practice self-regulation, 135, 193
Practice Standard, 15, 43, 75, 95, 102, 149–50, 159, 237, 239
precision, 9, 13, 161–63, 165–69, 171–78, 181–83
precision in mathematical skills, 166
predictions, 11, 134, 138, 196
preparation, 43, 77, 183, 185
preparing students, 241
prevention, 2, 6
Prioritizing intrapersonal skills, 10, 191
problem card, 59–61
problem card decontextualizes, 59
problem context, 162–63
problem formulation, 105
Problem identification and problem formulation, 105
problems, 21–30, 35–36, 38–43, 45–49, 56–62, 70–76, 105–7, 109–11, 113–21, 133–38, 156–58, 165–69, 171–72, 180–82, 188, 199–200, 213–14, 216–17, 219–20, 222–23
  addition, 71, 216, 222, 229
  approach, 121, 222
  authentic, 116
  breaking down big-picture, 67
  contextual, 64, 164
  decontextualize, 49
  design, 105
  difficult, 31, 34
  everyday, 131
  geometric, 25
  hard, 214
  individual, 188
  multiple, 217, 236–37
  multi-step, 26
  natural numbers, 186
  new, 213
  original, 22
  real-life, 164, 181–82
  real-world, 46–48, 53, 109, 111, 121, 130–31
  real-world application, 110
  resolving, 111
  sandbox, 127
  single, 213
  social, 106
  sticker, 230
problem Self-regulation Adaptability, 201
problem situations, 45, 47, 61, 76, 167
problem solvers, effective, 130
problem-solving, 10, 21–43, 46–47, 64–65, 108, 111–12, 118, 121, 171, 191, 207
  collaborative, 173
  effective, 23, 211
  efficient, 137
  promoting, 194
problem-solving abilities, 173
problem-solving actions, 24
problem-solving activities, 167
problem-solving approaches, 27
problem-solving efforts, 178
problem-solving ideas, 80
problem-solving information organizer, 64–65
problem-solving methods, 199
problem-solving process, 11, 47, 54, 87, 138, 167, 169
  reflective, 57
problem-solving strategies, 62
  multiple, 199
problem-solving tasks, 34
problem-solving techniques, 21
problem students, 113
problem types, 51, 213
  general, 213
Procedural skill, 29, 170

INDEX

process, 9, 18, 20–21, 47–48, 56–57, 74, 76, 106–7, 134, 152, 154, 173, 197–98, 214, 218–20
process demands time, 53
process information, 169
process standards, 15, 46
product, 47, 51–52, 189, 203, 223
proficient students start, 22
progress, 14, 17, 35–36, 67, 93, 95, 122–23, 148–50, 173, 175, 177, 194, 197, 201, 225–27
Promote Adaptability, 61
promote adaptability in students, 56
properties, 46, 87, 136, 157, 222–23
Protective factors, 4
prove, 72, 92, 214–15
providing constructive feedback to students, 167
purpose, 5, 8, 29, 106, 118, 135–36, 141–42, 165, 233, 240
puzzle, 63, 115–16, 217

quadrants, 208
quadratic equations, 82–83, 87–88, 99–100
   procedural fluency solving, 82
   students solving, 88
quadrilaterals, 155, 164, 178–80
quantities, 45–46, 54, 57–59, 61, 70, 73, 75–76, 161–62, 165, 172, 175, 196–97, 216
questions, 58–60, 70, 76–77, 81–82, 84–85, 87–88, 90–91, 102–3, 112–15, 117–20, 127–31, 133, 146–47, 155, 157–58, 200–202, 209–11, 219, 235–37, 241–42

Rating, 29
rational functions, 208, 210–11
rational numbers, 164, 180, 183, 187
   negative, 164, 180, 183
   nonnegative, 48, 51
real-world problems and abstract thinking and reasoning, 53
real-world problem-solving, 45
real-world situations, 52, 76, 106–7, 116, 127
reason, 45, 47, 52, 54, 67–68, 76, 80, 89, 109, 113–14, 120
reasonableness, 49, 135, 147, 164, 180, 214
Reason Abstractly, 46, 66

reasoning, 45–47, 53, 55–56, 79–81, 83–85, 87–88, 90, 95, 102–3, 121, 161–63, 166–67, 171–75, 193–95, 213
   quantitative, 45, 47, 62, 69, 75, 77
   relational, 196
reasoning and strategies, 47
reasoning processes, 28, 83, 220
recontextualize, 53, 107
record evidence, 122, 176
recreate, 41–42
rectangles, 21, 51, 87, 93, 108, 125, 136, 155, 164, 178–79, 220
referents, 45
Reflecting, 32, 34, 109, 173, 175
reflection, 15–16, 18, 24, 27, 42, 75, 84, 102, 174, 182, 210
Reflective practice, 20, 42, 182
reflective questions, 20
reflective thinking, 9, 54
refocuses, 71, 74
regularities, 22, 213–14, 217–18, 221, 223, 232, 237
regular reflection on emotions and actions, 69
relationships, 5, 8, 10, 22, 25, 45–46, 62, 64, 106, 115–16, 164–65, 185–86, 215–17
   repetitive, 237
relationships and connections, 5, 217
relationship skills, 86, 241
remaking tables, 235
remind students, 223
repeated reasoning, 214–37
repeated reasoning in mathematical problems, 220
repetitions, 215–17, 226, 237, 242
representations, 23, 25, 47, 62, 64–65, 86–87, 106–7, 165–66, 182, 186, 190, 198, 204
   mathematical, 47, 65, 106
   multiple, 25, 61, 64, 66
rereads, 70, 73, 75, 92
resolution process, 105
resources, 156–57, 241
Resources for student support, 3
responses, 10, 62, 75, 90, 141–42, 174, 191, 202, 207, 239, 241–42
responsibility, 79, 113, 139–40
responsible decision-making, 108, 112, 124, 130, 138

responsible decisions, 120, 138, 142
Revisit Table, 76, 102
rhombuses, 164, 178-79
right answers, 2, 147, 204, 218
risks, comfortable taking, 218
routine, 54-55, 57-59, 69, 71, 73, 88-89, 102, 118, 168, 192-93, 196-200
routines support, 54, 192
routines support students, 197
routines to teach students to pause, 88
routine tasks, completing, 188
rows, 38, 51, 206
rule, 196, 213-14, 220, 223-26, 235-37
  general, 217-18
rule keeper, 224
rulers, 41, 133, 135-36, 143-45, 147, 151-52, 154, 232

sample student exchange, 59
sandboxes, 125-27
scale, 25, 40-42, 116, 165
  involving, 25, 40, 42
scale factor, 25, 40-42
  new, 41
scenarios, 121-22, 175, 228
  everyday problem-solving, 22
schools, 1-4, 7, 90, 151, 241-42
screen, 3, 155-56
secondary example, 38, 69, 125, 151, 178, 199, 203, 227
second-grade class, 69, 151
SECs. *See* social-emotional competencies
SECs students, 242
SEL. *See* social-emotional learning
selecting mathematical standards and goals, 189
selection of tools, 143
self-assessment checklist, 37, 68, 96, 124, 151, 178, 203, 225, 227
self-awareness, 8-10, 13, 81, 83-84, 87, 89, 91-92, 94, 96-97, 99, 101-2
self-awareness and social awareness, 89, 92, 99
self-control, student's, 191
self-efficacy, 9, 13, 24, 27-29, 31-32, 34-40, 42, 55, 161-83, 221, 225
self-efficacy and interpersonal skills, 39
self-efficacy and social awareness, 34

self-efficacy in social-emotional competencies, 68, 202
self-reflection, 18, 36, 68, 91, 202
self-regulation, 45, 47, 53, 56, 58, 68-69, 72, 75-77, 165-66, 176, 178, 182-83, 191-92, 199-204, 210-11
self-regulation and focus, 180
self-regulation and self-efficacy, 183
self-regulation and sustained attention, 58
self-regulation skills, 57, 170-71
sense, 22-23, 25, 28-30, 35-36, 38-40, 42, 45-46, 74, 76, 80-81, 84, 100, 102, 106-7, 239
Sense-making, 37
set, 61, 73-74, 84, 115-16, 165, 167, 207, 216, 232, 236, 240
shapes, 16, 27, 87, 93, 115-16, 163-64, 178-80, 185, 198
Sharp monitors students, 228
shifting goals of education, 1-2
shortcuts, 146, 213-15, 217-20, 226, 236-37
sign, equal, 161-62, 164-65
situations, 9-10, 45-48, 54, 57-59, 65-66, 72-74, 76, 105-9, 111, 122, 137-38, 147, 154, 191-92, 228-31
  everyday, 79, 121, 228
  new, 59, 158, 188
skill building activities, concrete, 31
skill for self-regulation, 197
skill level, 145
skills, 6-7, 10-12, 14-17, 26-27, 43-44, 52-53, 76-77, 79, 85-86, 92-94, 124-25, 136-39, 158, 165-67, 171-73, 183-84, 191, 217, 240, 242
  21st-century, 6, 148
  brainstorming, 85
  cognitive, 10
  emotional, 241-42
  employability, 6, 241
  foundational, 110
  implied, 18
  life, 47, 120
  mathematical, 9, 166
  metacognitive, 79
  new, 1, 112
  possible, 85
  problem-solving, 21, 35, 167, 183
  social, 6, 12, 17, 217

INDEX

soft, 6, 241
technical, 140
skills and outcomes, 6
skill sets, 10, 135
skills in integrity, 158
skills of creative thinking and adaptability, 115
skills of empathy and assertiveness, 85
skills of self-awareness and social awareness, 92
skills support, 242
slope, 190, 195, 214–15, 232–33, 235
social, 6–7, 11, 19, 99, 176–77
social awareness, 24, 27, 29, 40, 42–43, 79–103, 108, 111–13, 120–22, 124–25, 127, 130, 165–66, 172–73, 176–78
practiced, 42
social awareness and communication, 173
Social awareness and relationship skills, 86
social awareness skills, 85, 122, 168
social awareness to construct viable arguments and critique, 85
Social-Emotional, 37, 68, 96, 124, 150–51, 177–78, 202–3, 226–27
social-emotional competencies (SECs), 4–7, 12, 14–20, 24, 34–35, 42–44, 67–69, 75–77, 93–94, 102, 137, 162–63, 175–76, 182–84, 191–92, 196–99, 202–3, 210–11, 214–17, 239–42
social-emotional competencies and enhanced lesson reflection, 18
social-emotional competencies development, 237
social-emotional competencies of self-regulation, 76
social-emotional competencies to maximize student outcomes, 4
social-emotional competency development, 52
social-emotional competency skills, 108
social-emotional development, 1, 4, 17, 24, 28, 53, 84, 113, 138, 166, 192
social-emotional goals, 17, 192
social-emotional learning (SEL), 4, 6, 30, 189, 220, 243
social-emotional learning competencies, 212
social-emotional learning perspective, 163
social-emotional skill development, 5, 18

social-emotional skills, 1, 4, 15, 24, 36, 56, 77, 136, 189, 211, 240–41
social-emotional support, 1, 3
social interactions, 11, 21, 40, 53
solution method, 88
solution pathway, 22–23
solution process, 83
solutions, 22–23, 26, 41–42, 46, 48–49, 59–60, 62, 64–65, 70–72, 101, 109, 112–13, 117, 127, 188, 217–18
incorrect, 72
potential, 105, 119
solutions and strategies, 41
solve word problems in situations, 137
solving equations, 48, 50, 196
solving mathematics problems, 57
solving multi-step problems, 183
solving problems, 21, 23, 87, 106, 112, 120, 185, 188
solving word problems, 130
Sophie Thinks She Can't, 220
sounds, 24, 38, 130, 139–40, 142
source, 29, 50, 52, 55, 60, 63, 110, 116, 118–20, 170–73, 175
Sources for Estimation Examples, 56
Sources for Video Versions of Word Problems, 120
spatial adaptability, better, 116
splash park, 69–71
Splash Park Task, 69–70
Sports, 73–74
sports equipment, new, 73–74
sports equipment set, 73–74
sprinklers, 69–71
square centimeters, 51
square root, 100–101
square roots, taking, 82–83, 88, 99
squares, 82–83, 87–88, 93, 115, 164, 178–79, 185, 221, 233
standard focus, 151
standards, 12, 15–16, 25–26, 48, 83, 95–96, 109, 111, 123, 136–37, 149–51, 190, 201, 210–11, 226–27
modeling, 111
start, 126, 128, 136, 139, 144, 215, 225, 227, 229, 231–32, 240–41
statement, 91, 103, 221
statistics, 111

steps, 9, 26, 90, 108, 110, 114, 122, 156, 167–69, 171, 173–74, 185–86, 212–13, 219, 234
steps support, 174
stickers, 228–31
   packs of, 228–29
sticky, 77, 212
story, 35, 54, 58, 70, 119, 122, 220
story problems, 48
   standard, 60
story/problem/situation, 58
strategic use of tools, 139, 141, 143, 147, 158
strategies, 9–10, 21, 23–29, 47–48, 53–54, 59–60, 72, 81, 83–84, 166–67, 169–70, 184–85, 188, 192–93, 196–97, 199–200, 215–16, 218–20, 229, 236
   appropriate, 188
   estimation, 49, 164, 180
   instructional, 17, 159, 237
   key, 21
   mathematical, 21
   multiple, 23
   sharing, 10, 42
   student's, 199
strategies and problem-solving methods, 199
strategies and routines support, 192
strategies and solutions, 59, 188
strategies for finding solutions, 29
strategies to make sense of patterns, 219
strengthen, 53, 188, 242
strengthen students, 44, 184
stress, 4, 9, 241
string, 51
structure and patterns, 193, 210
structure and patterns in mathematics, 193
structure in mathematics and management of emotions, 211
structure Lesson Objective, 13
structure of expressions, 199
structure of mathematics, 194
structure of problems, 214
structures, 3, 14, 17, 83–84, 101–2, 113, 127, 130, 185–86, 188–94, 196–203, 205–6, 208–12, 214, 218–19
   conceptual, 106
   equation's, 100

hybrid learning, 3
   mathematical, 191–92, 211, 214
structures and discern patterns, 192
structures to support students, 75, 102
student actions, 122, 175
   affirming, 200
student activities, 28
student awareness, 18, 176
student complaining, 127
student curiosity, 127
student development, 4
student discourse, 199
   routine leverages, 64
student engagement, 1, 31, 80
student exploration, 147
student justification, 79, 147
student learning, 43, 77, 183
Student Log to Record Response to Feedback, 34
student misconduct, 3
student progress, 176
   monitors, 74
student reflection, 69, 176, 202–3
student reflection of intrapersonal and interpersonal skills, 202
students, next, 207
student's ability, 47
students and demands on schools, 1–2
students approach tasks, 24
students brainstorm ideas, 112
students chime, 154
students contextualize, 54
Students' curiosity, 236
students decontextualize, 47
Student Self-Assessment Quick Check, 203
students examples, 181
students experience addition, 25
students experiment, 115, 229
students feedback, 77
students focus, 153, 181, 188
students for estimates, 55
students form, 192
student's idea, 200
students in brainstorming skills, 85
students in groups, 228
students in linking mathematics content, 24
students in managing behaviors and emotions, 192

INDEX

students in MP1, 38
students in MP2, 69, 72
students in MP3, 97
students in MP4, 125
students in MP5, 143
students in MP6, 164
students in problem-solving, 42
students in problem-solving activities, 167
students in strategic use of tools, 141
students instruction, 242
students leverage, 85, 138
students make sense of concepts, 107
students make sense of problems, 23
students make sense of quantities, 45
students MP1, 40
students pair, 73
students persevere, 23, 44
student's perspective/thinking/
    reasoning, 212
students plan, 25
students practice self-regulation, 175
students share connections, 206
students step, 211
students time, 58, 127, 206
students to attend to precision, 167
students to groups, 97
students to make, 128, 172
students to persevere, 34, 221, 242
students to plan, 188
students to step, 212
students to use structure of functions, 190
students transition, 61
students trust, 239
student success, 5
student support, 3
student thinking, 28, 146, 236
    harness, 51
student voices, 209
Student Work Sample, 234
subset, 164
subtraction, 24–26, 32, 38–39, 48–51, 72, 109, 164, 215, 228
subtraction problems, 26, 32, 48–49, 135
subtraction situations, 25, 49
success, 5, 8–10, 47, 145, 174, 242–43
    academic, 2, 135, 161
sugar cubes task, 119–20
sum of fractions, 82, 96

support, 4–5, 16–18, 26, 28, 77, 84, 140–41, 147–48, 165–66, 170–71, 175–76, 191–92, 217–18, 220–21, 240
support adaptability, 211–12
support appropriate and strategic use of tools, 147
supported students, 71, 154, 180–81
support for social-emotional learning, 220
support intrapersonal, 75, 77
support intrapersonal communication, 77
support mathematics engagement, 148
support self-regulation, 77, 211–12
support student interaction, 43
support students, 5, 24, 39, 43, 64, 75, 88, 102, 130, 158, 167
support student's ability to discern patterns and structure, 210
support students in making sense of problems and persevering, 43
support student thinking, 204
sustained attention, 13, 45–77
symbolic notation, 162, 168
symbols, 11, 26, 45–46, 48, 106, 109, 161–63, 165, 167, 169, 174–75

table, 8–11, 22–23, 33–34, 36–37, 49–51, 58, 66–68, 80–81, 85–86, 92–96, 122–24, 128–29, 133–35, 137–42, 147–53, 175–78, 193–95, 201–3, 224–29, 232–35
table equation, 66
table group, 228
table of benefits and limitations, 153
table of values, 137
tallies, 229
tangrams, 115
tasks, 27, 29–31, 53, 55, 69, 73, 82, 87–88, 93, 133–34, 141–46, 166, 173–75, 197, 236–37
    complete, 70, 140
    modeling, 117
    open, 77
tasks students, 181
tax, 73–74, 180–81
    sales, 180–81
teach, 1, 6, 12, 17, 20, 26, 29, 52–53, 115, 165, 167
Teacher Actions, 175
teacher and student activities, 28

teachers, 1–7, 17–18, 24, 42–43, 46, 58–59, 69, 71–72, 75, 89–90, 129–30, 145–46, 157–58, 175, 182–83, 193, 199–202, 210, 220–21, 242
teachers guide students, 211
teaching, 1, 4–6, 12, 15, 19–20, 27–28, 77, 81, 180, 183, 236, 240, 242
teaching students, 1, 82, 99, 139, 167, 170, 219
teach self-regulation skills, 56
teach students, 88, 147, 167
team, 218
teamwork, 10–11, 14, 215, 218, 220–21, 224–27, 235, 242
technology, 90, 133–35, 138–39, 159, 189
technology and tools, 138
template, 89–90
    robust lesson plan, 12
terms, 6, 137, 164–65, 169, 173–74, 181, 185, 187, 214, 216, 233–34
thinking, 39, 72, 74, 87–88, 99–100, 116, 124, 192–93, 195–200, 206–7, 209–10, 212, 219–20, 222–24, 242
    abstract, 53–54
    deeper, 31, 206
    mathematical, 51, 88, 167, 182, 236
    strategic, 84, 196
    student's, 205
thinking process, 28, 75
thinking skills, higher-order, 42
third-grade students, 164
Thompson, 72–75
Three-Act Math Tasks, 118, 131
time, 2–4, 6, 9–10, 25–27, 31–34, 52, 59, 71–73, 152–54, 171, 188–89, 192–93, 197–99, 206–9, 212–13, 216–18, 223, 230–31, 234–36, 242
    best, 89
    costs, 52
    dedicate, 162
    extended, 2
    involving, 26
    quiet, 128
    short, 74
time control variables, 127
time for math, 2
time intervals, 26
timer, 56, 129

tip, 180–81
tool for completing routine tasks, 188
tools, 9, 14, 17, 21, 23, 28, 84, 86–87, 106–7, 133–59, 164, 166, 168, 192
    appropriate, 13, 107, 133, 136, 139, 151
    available, 133, 138
    avoiding, 148
    changed, 146
    complex, 203
    effective, 64
    efficient, 133, 143, 158
    instructional, 43, 131
    misusing, 140
    monitoring, 236
    new, 147
    nonstandard measuring, 154
    observation, 68, 149, 202
    right, 133
    self-assessment, 94, 227
    self-reflection, 201
    technological, 134–35
tools and strategies, 9
tools and structures to make, 127
tools for classroom and life, 159
tools for efficient problem-solving, 137
tools for students, 146
tools students, 152
tools to measure, 136
tools to support, 140
tools work, 146
Traditional word problems, 28, 131
transforming a mathematical problem, 110
triangles, 136, 155, 190, 215, 232, 235
    small, 115
type, 83, 90, 127–28, 148, 154, 186, 190, 214, 219–20, 222–24, 228, 237, 242
    new problem, 188

ultimate learning skill, 117
Uncover Mathematics Equation, 197
understanding, 8, 21–23, 25, 27, 50–51, 56–57, 71–72, 83–84, 105–6, 130–31, 136, 147–48, 165–66, 175–76, 180–81, 194–96, 198
    conceptual, 28–29, 51, 64, 161, 215, 237
    procedural, 139
    structural, 189
understanding of addition and subtraction, 39

INDEX

understanding of emotions and actions, 176
understanding of equations, 51
understanding of intrapersonal and interpersonal skills, 42, 182
units, 46, 149, 167, 169, 175, 194, 232
Unknown, 26, 48, 50–52, 109
Use, 30, 36, 94, 97, 133, 135, 139, 147, 151, 175–76, 185–86, 189–90, 203
Use addition and subtraction, 25, 109
use appropriate tools, 133
Use observation tools, 159, 237
Use of Tools, 145–46
use structure, 185–211
Use Table, 36

value of intrapersonal and interpersonal skills, 43
values, 9, 43–44, 84, 93, 106, 137, 139–40, 184–86, 189–90, 196, 204
    initial, 137
variables, 107, 121, 128, 130, 165, 186, 196
varied engagement strategies, 158
video, 118–19
Video Story Problems, 121
vignettes, 38, 154, 178
visual fraction models, 82, 165

vocabulary, 139, 162, 168
    precise, 174–75

watch for evidence of student actions, 122
Well-Rounded Math Lesson Guide, 12–13
wonder, 71, 91, 117–19, 129, 201, 205–6, 217–19, 230–32
Wonder Routines, 88, 117–18, 203, 228, 232, 236
word problem in stages, 57
word problems, 25–26, 48, 51, 57, 59, 87, 109–11, 120, 137, 164, 187
    multi-step, 48
    real-world, 117
    solve, 109
    students decontextualize, 109
    two-step, 48, 76
words, 46, 55, 70, 73, 79, 99, 107, 110–11, 137, 139, 205

x-axis, 233

y-axis, 233
Younger students, 22

zeros, 189–90, 193, 203–5, 211

CORWIN Mathematics

Supporting TEACHERS | Empowering STUDENTS

PETER LILJEDAHL
Fourteen optimal practices for thinking that create an ideal setting for deep mathematics learning to occur.
Grades K–12

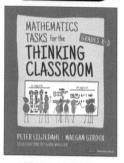

GRADES 6–12 COMING SOON!

PETER LILJEDAHL, MAEGAN GIROUX
Delve deeper into the implementation of the fourteen practices from *Building Thinking Classrooms in Mathematics* by focusing on the practice through the lens of tasks.
Grades K–5

FRANCIS (SKIP) FENNELL, BETH McCORD KOBETT, JONATHAN A. WRAY
Leverage five formative assessment techniques that lead to greater attention to planning, stronger instruction for teachers, and better achievement for students
Grades K–12

BETH McCORD KOBETT, FRANCIS (SKIP) FENNELL, KAREN S. KARP, DELISE ANDREWS, LATRENDA KNIGHTEN, JEFF SHIH, DESIREE HARRISON, BARBARA ANN SWARTZ, SORSHA-MARIA T. MULROE, BARBARA J. DOUGHERTY, LINDA C. VENENCIANO
Detailed plans for helping students experience deep mathematical learning.
Grades K–1, 2–3, 4–5, 6–12

MAGGIE LEE McHUGH
Move beyond task-oriented mathematics to truly memorable and equitable project-based learning with the simple, research-based game plan mapped out in this approachable guide.
Grades K–12

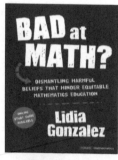

LIDIA GONZALEZ
Dismantle societies' pervasive beliefs and stereotypes about mathematics so we can reform math education to work for everyone.
Grades K–12

To order, visit corwin.com/math

JENNIFER M. BAY-WILLIAMS, JOHN J. SANGIOVANNI, ROSALBA SERRANO, SHERRI MARTINIE, JENNIFER SUH, C. DAVID WALTERS, SUSIE KATT

Because fluency is so much more than basic facts and algorithms.
Grades K–8

MARIA DEL ROSARIO ZAVALA, JULIA MARIA AGUIRRE

Discover innovative equity-based culturally responsive mathematics instruction that unlocks the mathematical heart of each student.
Grades K–8

IVANNIA SOTO, THEODORE RUIZ SAGUN, MICHAEL BEIERSDORF

Focus on the literacy opportunities that multilingual students can achieve when language scaffolds are taught alongside rigorous math and science content.
Grades K–8

JOHN J. SANGIOVANNI, SUSIE KATT, LATRENDA D. KNIGHTEN, GEORGINA RIVERA, FREDERICK L. DILLON, AYANNA D. PERRY, ANDREA CHENG, JENNIFER OUTZS, KAREN MESMER, ENYA GRANDOS, KEVIN GANT, LAURA SHAFER

Actionable answers to your most pressing questions about teaching elementary math, secondary math, and secondary science.
Elementary, Secondary

CHRISTA JACKSON, KRISTIN L. COOK, SARAH B. BUSH, MARGARET MOHR-SCHROEDER, CATHRINE MAIORCA, THOMAS ROBERTS

Help educators create integrated STEM learning experiences that are inclusive for all students and allow them to experience STEM as scientists, innovators, mathematicians, creators, engineers, and technology experts!
Grades PreK–5 and Grades 6–12

CORWIN

CORWIN

To help every educator help every student

We believe   that every single student deserves a great education

We believe   that knowing our impact is both a privilege and a responsibility

We believe   that a fair, stable, and thriving society is built on education